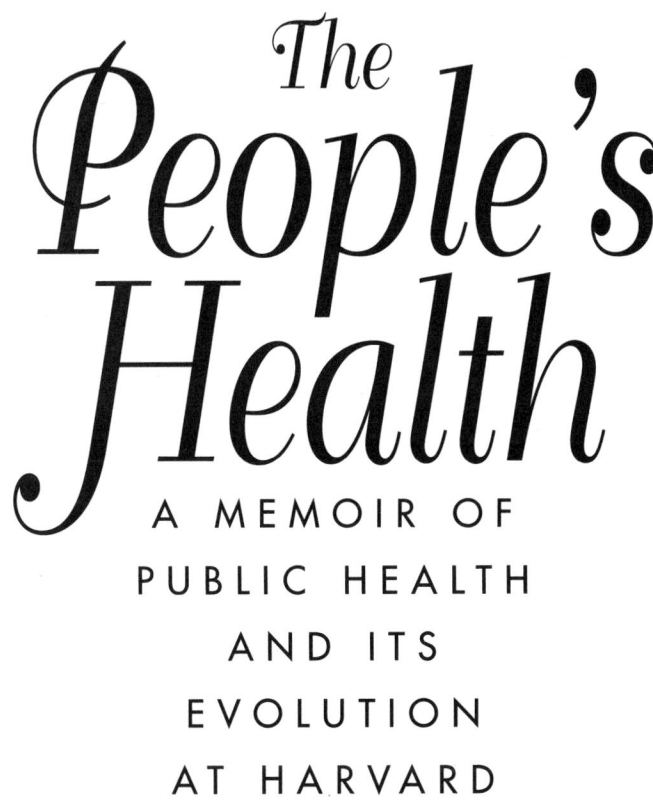

The People's Health

A MEMOIR OF PUBLIC HEALTH AND ITS EVOLUTION AT HARVARD

ROBIN MARANTZ HENIG

JOSEPH HENRY PRESS
Washington, D.C.   1997

JOSEPH HENRY PRESS • **2101 Constitution Ave., N.W.** • **Washington, DC 20418**

The Joseph Henry Press, an imprint of the National Academy Press, was created with the goal of making books on science, technology, and health more widely available to professionals and the public. Joseph Henry was one of the founders of the National Academy of Sciences and a leader of early American science.

Library of Congress Cataloging-in-Publication Data

Henig, Robin Marantz.
    The people's health: a memoir of public health and its evolution
at Harvard / Robin Marantz Henig.
       p. cm.
    Includes bibliographical references.
    ISBN 0-309-05492-3 (alk. paper)
    1. Public health—History. 2. Harvard School of Public Health—
History. I. Title.
RA424.H46   1997
362.1'09—dc20   96-41963
                                               CIP

Printed in the United States of America

Dedicated to the truly dedicated:
the public health pioneers

Figures do not measure the terror of epidemics,
nor the tears of the mother at her baby's grave,
nor the sorrows of the widow whose helpmate
has been snatched away in the prime of life.
To have prevented these not once, but a million times,
justifies our half century of public health work.

— Charles V. Chapin (1856–1935)
Rhode Island Commissioner of Public Health

Foreword

To train enough professionals to keep all the people well by curing disease would involve an expenditure of so large a fraction of the income of any nation as to make the cure less attractive to the taxpayers than the disease. Indeed, if we tried to do the job today without our present knowledge of . . . public health, the situation would be hopeless. A layman needs only a moment's reflection on the story of the control of typhoid fever and smallpox to be convinced of that. . . . [A]s the profession of public health advances, the need for hospitals, physicians, and surgeons will diminish. The corresponding costs of keeping people well will decrease and the entire people of a free and prosperous society will enjoy a state of health undreamed of even by the kings and nobles and the privileged classes of a few centuries ago.

— James Bryant Conant
(president of Harvard University, 1933–1953),
Public Health in the World Today

This declaration of purpose is as timely today as when it was written nearly 50 years ago. At that time the Harvard School of Public Health was just over a quarter-century old and celebrating its new status as a freestanding faculty of Harvard University—on a par with each of the professional schools and no longer a part of Harvard Medical School. The book for which Mr. Conant, then Harvard's president, provided the foreword was written by and for experts in public health. The present book, produced on the occasion of the school's seventy-fifth anniversary, has been written instead for the lay reader, reflecting the belief that it has never been more important to foster public understanding of what is being done and can be done to preserve and advance the people's health.

This volume shines a light on the history of public health as refracted through the lens of the Harvard School of Public Health. Other schools, organizations, and agencies can, of course, also boast of numerous accomplishments that have contributed to the success of public health over the years. This book, organized thematically and highlighting the work of Harvard faculty and graduates, is less a history per se than an institutional memoir. Any 75 year old, human or institution, might be permitted the indulgence of examining how one's life has affected and been affected by the events and advances of the previous decades.

In this book you will read about many pioneers in the field of public health, past and present, whose exploits convey the excitement of learning and discovery. These explorers armed themselves with a vision of a healthier world. As they confronted such dangers as typhus, malaria, tuberculosis, and polio, they extended the boundaries of knowledge and crafted the tools that have led us closer to bringing such killers to bay. Today's health explorers examine the causes of devastating illnesses such as heart disease, cancer, and AIDS, seeking ways to prevent them. Others pursue a deeper understanding of the fundamental connections among social forces, personal choices, and health.

Inventors, sleuths, advocates, scientists—these are terms that describe public health professionals and researchers. For many, public health is a cause; for others, it is a fascinating array of thorny problems begging for analysis and resolution. While public health professionals invoke the different languages of specialized disciplines, they are joined in a common purpose—to defeat diseases and other risks that threaten populations. Thanks to their achievements, children in many countries can grow up free of polio, rubella, deadly diarrhea, and lead poisoning, and they are protected from injury, safely helmeted on bicycles or seatbelted in the family car.

Despite these advances, many health threats persist. Pollution fouls the world's atmosphere, subtracting healthy years from the lives of adults and children. The increasing use of tobacco in developing countries threatens women and men with cancer and premature death. The resurgence of such diseases as tuberculosis and drug-resistant malaria confounds predictions and dispels complacency borne of yesterday's successes. Today, global commerce and travel provide a virtual freeway around the world for the spread of diseases both old and new. Without a broad understanding of health, its relation to economic development, and its connection to social change and human rights, we cannot hope to thrive as a world community. In the coming decades, public health can function as a gateway to peace because it offers the world a vocabulary, science, and methods for meeting the most basic needs of every citizen.

This book brings to life the story of public health, thanks to the skills of author Robin Marantz Henig. These pages are animated by the many Harvard faculty, alumni, and others who served as resources as she sought to convey the many facets of public health. We could never begin to tell all the stories and highlight all the discoveries in our wide-ranging field. It is hoped that by having chosen some of the salient ones, you, the reader, will share in our enthusiasm and join us in the great cause of making the world a healthier place for all people.

To faculty, students, and alumni of the Harvard School of Public Health, my profound admiration for your energy and your commitment in carrying the mission of the school to every corner of the globe. To the administrators and staff of our school, my heartfelt thanks for your dedication and hard work. To friends of the school, my deepest gratitude for your vision and your support. As we pay homage to the triumphs of the past and present, let us also rededicate ourselves to the task ahead.

Harvey V. Fineberg, M.D., Ph.D.
Dean, Harvard School of Public Health
Boston

Foreword

Contents

The People's Health

The Mission of
Public Health

Public health experts love to tell the story of John Snow and the Broad Street pump. At the midway point of the nineteenth century, cholera was one of the deadliest diseases in Europe. It wiped out whole communities in a matter of weeks, and no one could determine how it was spread—or how it could be stopped.[1]

John Snow applied logic to a problem that had previously caused panic and paralyzed health workers. Snow was a British physician of such renown that he had been called in as an anesthetist when Queen Victoria delivered her first child in 1853. Approaching the cholera question, he sat down one afternoon in 1854 with a map of London, where a recent outbreak had killed more than 500 people in one dreadful 10-day period.

He marked the locations of the homes of those who had died. From the marks on his map, Snow could see that the deaths had all occurred in the so-called Golden Square area. The most striking difference between this district and the rest of London was the source of its drinking water. The private water company supplying the Golden Square neighborhood—which, according to the tradition of the day, was different from the private water companies supplying other neighborhoods—was getting its water from a section of the Thames River that was known to be especially polluted.

So Snow went down to Broad Street, where he suspected that one particular pump was the source of the contaminated water. And, in a gesture that still reverberates among public health scholars today, he removed the handle of the Broad Street pump.

Once the pump was out of commission, the epidemic abated. Snow did not know—nor did anyone else in that era—exactly how cholera had been passed from one person to another. Indeed, no one was even to conceptualize the notion of germs as a cause of disease for nearly 30 years. But Snow had shown that the cholera poison, whatever it was, could be found in contaminated water and taken in by mouth. The precise nature of the poison, and its exact route of transmission, did not really matter to him. All he wanted to do was keep it from being ingested and passed on.

John Snow is a hero to public health experts because he was the first to conduct a careful epidemiological investigation and the first to take definitive action and get dramatic, clear-cut results. The basics of Snow's work are still required reading for many students of epidemiology today. His work was based on elements of scientific investigation that are central to public health today: measure the distribution of disease events in a population; define the problem, at the same time bringing together all the necessary experts; and design and implement an intervention. Snow's contributions go beyond the removal of the Broad Street pump; he devised the first large epidemiological study for understanding a disease problem in a population. This approach has led to some of the most dramatic public health "detective work" in the past half century—the conquest of smallpox, the treatment of river blindness, and growing understanding of AIDS.

Despite this string of victories, most people do not know exactly what public health is. They confuse it with any undertaking involving disease, medicine, or health care. In fact, from a public health perspective, medicine is really one among many instruments of public health. Public education, better nutrition, environmental protections, workplace safety, and medical care are strategies to advance the public's health. Today, public health researchers include policy analysts, medical experts, and economists who strive to make medical care more affordable, widely available, and of the highest quality. At the same time, other public health professionals are working to enable more healthy lifestyle choices and to promote healthier physical and social environments.

The key difference between medicine and public health is in the way public health professionals, as opposed to clinical practitioners, look at a health problem. Faced with an illness, a personal physician will try to relieve an individual patient, while a public health professional will examine an entire population and develop interventions with the goal of preventing similar illness in the future.

"The question I am asked more than any other is, 'What is public health and how is it different from medicine?'" says Harvey Fineberg, dean of the Harvard School of Public Health. "And what it boils down to is that medicine has a different emphasis." The focus of a public health professional's gaze is always on a large cohort of people, whether it be an entire neighborhood, everyone in a certain ethnic minority, all children, all mothers, all old people, all pregnant women, or all industrial workers. "One emphasis is not necessarily better than the other," says Fineberg. "But in my heart of hearts, I really believe we will never in this world solve health problems one person at a time."[2]

At the turn of the century, a newborn baby girl had a life expectancy of 44 years; today, her life expectancy at birth is 78.[3] This improvement is not a medical victory; it is a public health victory. By the time doctors armed themselves with antibiotics and vaccines in the 1940s and 1950s, most of the advances against infectious disease already had been made by the sanitarians who cleaned up the filth of the cities and the countryside, ensured that the drinking water was clean and that wastes were safely disposed of, and limited the spread of waste-borne microorganisms.

As we will explore in the remainder of this book, many improvements in the public's health have occurred in the same way. Throughout, we will look at health issues from the public health perspective, that is, from the vantage point of health promotion and disease prevention in whole populations. In this introductory chapter, which lays out the book's structure and overriding themes, the goal is to answer for the general reader that first and most important question: What is public health?

In the early years of public health, a century or more ago, the profession tended to confine itself to engineering interventions. Public health meant sewer development, sanitation, and purification of food and water. Gradually, the profession branched out, to encompass a variety of disciplines, usually in response to scientific developments that allowed for more effective disease prevention techniques: pasteurization of milk; control of cholera and yellow fever; vaccinations against diphtheria, tetanus, whooping cough, and smallpox; reduction of hazards in the workplace. In each endeavor, John Snow's essential method was repeated and refined. The mantra of public health was then, and still is, to measure the distribution of the problem, define the causes, and design and implement an intervention.

Shortly after World War II the profession of public health—like many other scientific disciplines—underwent a profound intellectual shift. Modern public health professionals no longer saw their mission in stark or simple terms. "Public health today is revealed as no mere matter of sewage systems, inoculations and reporting of children's diseases," wrote the editors of *The New England Journal of Medicine* in 1951. The field was facing new challenges that required, as they put it, "a perceptible reorientation, a resynthesis of public-health teaching, practice and research."[4]

That reorientation inevitably expanded the stated mission of public health. Public health's goal was to make possible the highest state of health for the greatest number, most prominently by ensuring the purity of whatever people ate, drank, and breathed. When public health was brought into academia, a new wave of organized intellectual effort burst forth: the critical problems and methods were defined and people

3

formally trained in these methods. The pragmatic "detective work" of the earlier years by pioneers out in the field was augmented by a deliberate effort to develop rational, objective study and investigation. In this way it was expected that faster progress would be made.

Bolstered by the effectiveness of its most basic tools—epidemiology and biostatistics—the modern mission of public health is to ensure health and reduce premature deaths from such disparate causes as drug and alcohol abuse, unintentional injuries, violence, poverty, human rights abuses, and lack of access to health care. To meet this goal, public health

William Foege

professionals have had to confront not only pathogens but also a huge range of other factors that affect the public's health: race, income, genetic predispositions, social infrastructure, even a population's belief in its own collective future. As a result, public health professionals have had to become conversant not only with the basic biomedical sciences but also with, among other fields, economics, ethics, politics, psychology, and sociology.

Such an interdisciplinary approach has stretched the organizational categories traditionally used by both schools of public health and community-based health departments. Indeed, today some people see the boundaries of public health as almost without limit. "Our job in public health is to be indignant on behalf of everyone," says William Foege of the Carter Center in Atlanta. "Every problem is a public health problem."[5] Said another way, public health today embraces so many difficult problems because the public health approach, as will be illustrated in this book, offers promising prospects for solutions where others have failed. Whether because of the threat of heart disease, adolescent violence, or air pollution, society increasingly turns to public health to define the problem and develop solutions because, too often, efforts by legislators, the medical community, the business sector, or law enforcement, acting alone, have fallen short.

A History of Good Results

A field that takes as its domain virtually all of humankind's health problems certainly has its work cut out for it. Fortunately, throughout

4

its long history, public health has managed to come through with some rather dramatic victories, even if in the early years some of them were pure happenstance. As Harvey Fineberg explains, "we begin with the problem, we measure it, and then we mobilize the methods and the people needed to solve that problem."[6]

This pragmatic approach has ancient roots. For example, the early Hindus thought demons caused disease, in particular the most deadly disease of the Indian subcontinent, a chills-and-fever illness known as the ague. The ague was thought to be caused by a demon who disguised himself as a mosquito, and people were taught to avoid the demon mosquito to protect their lives. Many of the symptoms of the ague were identical to what is now called malaria, which is of course transmitted by mosquitoes. In retrospect, then, the Hindu admonition to stay away from mosquitoes, though based in animistic theology, turned out to be of solid public health value.

"In everyday life," note two historians who observed this phenomenon, "it mattered little whether one regarded the mosquito as a demon, or as an insect bearing in its belly malaria-producing germs—provided one avoided the mosquito."[7]

Animism was also at work in Manchuria, where Asian nomads believed that departed ancestors were often reincarnated as marmots (a kind of rodent related to a woodchuck or prairie dog). Their religion required them to treat these animals with special care:

> Trapping was taboo; a marmot could only be shot. An animal that moved sluggishly was untouchable, and if a marmot colony showed signs of sickness, custom required the human community to strike its tents and move away to avoid bad luck.[8]

Such conventions were of great benefit to the nomadic tribesmen living in the steppes of Manchuria, because they kept the people away from the rodents most likely to harbor the bacillus that causes bubonic plague.[9]

Even modern public health triumphs still work without their champions quite knowing why. Today, public health experts continue to face a group of conditions for which they can offer little more than post hoc explanations for apparent successes. Unlike the more familiar dangers, the transmission of certain infectious diseases, nutritional deficiencies, and cigarette smoking, where progress in understanding the mechanisms of how infection or harm occurs has been relatively steady, other threats to health are so complex that we still do not understand their basic causes, much less the tactics needed to eradicate them. Occasionally, a problem such as handgun violence, teenage pregnancy, obesity, infant mortality, or overpopulation will be met with an effective public health approach—but practitioners might be nearly as unable to explain why those solutions worked as the ancient Hindus were to explain why it was wise to avoid the demon mosquitoes.

Some of the most dramatic early public health victories were actually achieved in the name of a totally misguided view of what made people sick and what kept them healthy. At the end of the nineteenth century, when the biggest hazard to public health was filth, cities were cleaned up by "miasmatics," who believed that disease was carried by putrid air rising from garbage, swamps, and dung. It was obvious to any careful observer that something needed to be done about the squalor of even relatively prosperous urban areas in the late nineteenth century. Living conditions at the time were filthy to such an extent that Americans today cannot even imagine, no matter how fetid the slums they have seen at home or abroad.

> Festering piles of garbage littered urban streets, dead animals lay where they fell, privies and cesspools overran into drainless, unpaved streets. Horses defecated indiscriminately.[10]

The large-scale cleanups mounted to eliminate dangerous "miasmas," though undertaken for the wrong reasons, served to eliminate the disease-causing bacteria and other microbes that were the true cause of disease outbreaks of the day. Public health officials at the turn of the century greatly reduced the spread of dysentery, typhoid, and other water-borne infections by insisting on "clean water and flushing away sewage in cast-iron pipes, . . . washing one's clothes and body regularly, and . . . stipulating that food and drink no longer be adulterated."[11]

Cleaning up the cities—
the sanitarians

But just because a problem was obvious and its solution straightforward does not mean cleaning up the environment was a simple matter. Even a century ago, much of public health was of necessity interdisciplinary. Think of all the different fields of expertise that came into play when cleaning up a city's supply of drinking water:

> Physicians diagnosed contagious diseases; sanitary engineers built water and sewage systems; vital statisticians provided quantitative measures of births and deaths; lawyers wrote sanitary codes and regulations; public health nurses provided care and advice to the sick in their homes; sanitary inspectors visited factories and markets to enforce compliance with public health ordinances; and administrators tried to organize everyone within the limits of their budgets.[12]

Because of the interdisciplinary nature of public health, unifying it into a single profession has meant bringing together all these different perspectives into what Elizabeth Fee of Johns Hopkins University calls "a single vision, a common philosophy."[13] That has been no easy task— and, as will be seen in the chapters to come, it is a task that has not always been accomplished successfully.

Sometimes, however, that "single vision" is magnificently orchestrated and expressed in surprisingly simple terms. At Harvard University, for instance, the strength of the "common philosophy" of public health is captured every year on commencement day. On a June morning in Harvard Yard, the president of the university confers degrees on graduates of every school and college. To each group of graduates he offers his own version of a mission statement, saying in a sentence or two what it is that these young people are now trained to do. When he awards degrees to new graduates of the university's school of public health, he tells the new public health professionals that they are now "ready to advance the welfare of peoples everywhere by the prevention of disease and promotion of health."[14]

This Book: The Text and Subtext

These are noble goals—the prevention of disease and promotion of health—and a description of the ways in which the profession has pursued them is the primary purpose of this book. Telling the stories of the leading public health accomplishments of the past half-century is the most vivid way to demonstrate how the profession has evolved. The evolution has been two pronged, encompassing both the overall mission of the field and the development of new tools, techniques, and expertise to meet that changing mission.

The primary purpose here, then, is to paint a history of public health with a broad-enough brush to describe major chapters of this story. The history of public health encompasses nearly every human endeavor, says Fee: "how populations experience health and illness, how social,

The founders—(left to right) George Whipple, William Sedgwick, and Milton Rosenau

economic, and political systems structure the possibilities for healthy and unhealthy lives, how societies create the preconditions for the production and transmission of disease, and how people, both as individuals and as social groups, attempt to promote their own health or avoid illness."[15]

But this book has a subtext as well. Its secondary goal is to fill in the broad brush strokes of the history with the shadings and highlights of one institution: the Harvard School of Public Health. Because while this book is at its core the story of an entire field, it is the story, too, of one particular place within that field.

The Harvard School of Public Health is the direct descendent of the first professional training program of public health in America, a joint venture between Harvard and the Massachusetts Institute of Technology known as the Harvard-MIT School for Health Officers.[16] The school was formed in 1913 in response to the scatter-shot training found among most professionals working in state and local health departments of the day. Milton J. Rosenau, professor of preventive medicine at Harvard Medical School and one of the initiators of the new school, explained the venture as follows:

> The country needs leaders in every community fitted to guide and
> instruct the people in the art of hygienic living; qualified to direct the
> expenditure of energy, time and money in public health work into
> fruitful channels; and able to initiate plans to meet novel conditions
> as they arise.[17]

8

Within a few years the Harvard-MIT school dropped its MIT affiliation. For one thing, the collaboration made it impossible for the School for Health Officers to grant degrees from either of the two sponsoring universities; all it could offer were certificates, which were not as prestigious. For another, a school under the complete auspices of only one university better fit the mold of a public health school as envisioned by one of the primary sponsors of such schools at the time, the Rockefeller Foundation—a sponsor that Harvard was hoping to enlist on its own behalf. Once the school split off from MIT in 1922 and was established as the Harvard School of Public Health, the Rockefeller Foundation did indeed provide a sizeable grant. At this point, Harvard's school joined the ranks of the two other leading institutions of the day, the Johns Hopkins University School of Hygiene and Public Health and the Yale University School of Medicine's Department of Public Health.[18] The Rockefeller Foundation meanwhile went on to support the establishment of public health institutes and training programs in dozens of countries.

Like the schools of public health at Hopkins and Yale, the school at Harvard began as a hybrid typical of some schools of public health even today: a quasi-independent professional school that, because of organizational lines of command, was as intertwined with the university's medical school as a pair of Siamese twins. But in 1946 that interdependence changed. The president of Harvard, James B. Conant, decided to separate the university's school of public health from its medical school—up until then the two schools had shared a single dean—and he appointed a public health dean who was to report directly to him. Conant's action finally gave the Harvard School of Public Health the status of an independent, degree-granting body comparable to any other school within the university. This allowed the research and training effort at the school to finally take off and to achieve parity with the other professional schools within one of the nation's premier universities.

Since World War II, the Harvard School of Public Health has had four deans: James Stevens Simmons (dean from 1946 to 1954), who had just retired from the U.S. Army as a brigadier general and who was a 1939 graduate of the Harvard School of Public Health; John Snyder (dean from 1954 to 1971), a bacteriologist who came to the Harvard School of Public Health in 1946 to pursue research into typhus and other rickettsial diseases, international health, and population control; Howard Hiatt (dean from 1972 to 1984), a Harvard-trained physician and chief of medicine at the Harvard-affiliated Beth Israel Hospital; and Harvey Fineberg (dean since 1984), a professor in the school's Department of Health Policy and Management and the first M.D. to receive a Ph.D. (in public policy) from Harvard's Kennedy School of Government.

As the school moves into its septuaquinennial year—that is, its seventy-fifth anniversary—this book is intended to serve as a kind of insti-

9

James Stephens Simmons

John C. Snyder

Howard H. Hiatt

Harvey V. Fineberg

tutional memoir: a selective and idiosyncratic picture of one of the nation's most productive and innovative schools of public health. Many of the changes that have occurred over the past half-century in public health have had their origins at Harvard; others have been reflected at Harvard through changes in its research agenda or curriculum. Harvard scientists were leaders, for instance, in conducting investigations on normal growth rates in children, which allowed physicians and parents to spot malnutrition with ease. Scientists at the Harvard School of Public Health, in collaboration with scientists at Harvard's Medical School, conducted landmark research into how to grow the polio virus in the laboratory, a first step on the road to one of the most dramatic public health discoveries of the century: the polio vaccine. More recently, they mounted long-term, large-scale population studies on lifestyle and disease to help clarify the health implications of food, exercise, and history of taking birth control pills or other medications. And Harvard was among the first schools of public health to include faculty members from nontraditional fields, primarily in the social sciences, both pioneering and reflecting the broadening of the field to include the exploration of such questions as injuries, violence, population growth, or Medicare reimbursement from a public health perspective.

The school's contribution to the teaching of public health has been no less important than its contribution to the knowledge base. "The educational mission of a school of public health is different from that of a medical school," says Marie McCormick of the Harvard School of Public Health. "Schools of public health must emphasize statistics, epidemiology, and the interplay between the biological and the social sciences. These analytic and interventional skills for disease control and eradication are not taught in medical schools; they are quite different from the skills needed to take care of sick patients."[19] Fineberg clarifies the unique educational goals of his school by drawing an analogy. "If a medical school is akin to a school of law," he says, "then a school of public health is like a school of justice."[20]

The coming chapters are guided organizationally by the principle that protection of the public's health requires the simultaneous operation of several perspectives on health and illness. For the sake of clarity, we will separate these perspectives and consider each one in turn. But this is an artificial separation, created primarily to give some shape to the order in which the book's stories unfold. First, focusing on the microbial pathogens that cause disease, we will look at the biological perspective. Then we will examine the physical perspective, looking at, for example, the ways in which health is affected by chemical additives and by the purity or impurity of air or water. Next comes the social perspective, an assessment of such factors as poverty, racism, violence, and politics. We will then look at the behavioral perspective, figuring in the choices that individuals make about their own lifestyles. Next, considering such issues as prevention, screening, and access to care, we will look at health and illness from the perspective of the health care system as a whole. Finally, from the vantage point of such subjects as human rights, population, and international crisis, we will focus on the global concerns of public health.

Despite the book's structure, the reader must remember that there are no comparable lines of demarcation in real life. While we have separated these perspectives in order to organize the stories we want to

11

tell, these are not mutually exclusive points of view. When confronting health conundrums in the real world, most public health officials are juggling several perspectives at once.

Three Stories

As we tell the tales of public health's most notable accomplishments, the careful reader will see a pattern emerge. Many of these stories begin with an emphasis on the biological or physical environment. These factors are tangible, relatively easy to measure, and easy to fight. But as the stories develop, other factors become salient; even the problems that seemed firmly rooted in, say, a mosquito-borne virus or a toxic air pollutant end up entangled with all manner of societal complications. In case after case the social environment in which a pathogen or toxin is transmitted becomes every bit as important as the original agent. To show how this works, we will consider here the stories of three important public health victories from the past century: the fight against hookworm, one of the major health concerns of the early 1900s; the fight against malaria, which heated up most dramatically during World War II; and the fight against PKU, a major cause of mental retardation in the 1950s and 1960s that has all but disappeared thanks to large-scale screening efforts. In each case the pattern is easy to spot.

Hookworm is a parasitic disease that causes anemia and chronic fatigue, leading to its epithet as the "germ of laziness." It is spread through the soles of the feet to people who step into soil contaminated with human excrement that contains worm larvae. Once they enter the body, the larvae make their way through the leg and into the small intestines. They hatch into half-inch-long adult worms with hook-like teeth (hence the name) that pass back into the soil or water in the individual's stool. A heavy infestation can involve 100 such worms at a time.[21]

In the early 1900s, hookworm was endemic throughout the southern United States. As conscientiously as public health officers tried to attack the disease—aided by $1 million provided by the Rockefeller Foundation's Sanitary Commission to Exterminate Hookworm Disease—their efforts were always hampered by the cold fact of poverty in the South. Hookworm was finally stopped only when everyone in the endemic area started to wear shoes.[22]

Malaria presented a similar quandary. By World War II, public health officials knew that it was passed by the *Anopheles* mosquito, and they had developed several techniques to limit the mosquito exposure of vulnerable populations: eliminating sources of standing water in which mosquito larvae bred and protecting people with mosquito nets, window screens, and other physical barriers during the times of day when mosquitoes were most likely to bite. Scientists knew how to prevent malaria with a medication known as Atabrine. But without a societal commitment to raising the general standard of living, there was no good way to take advantage of any of that biological knowledge. As historian John Duffy explains:

> Once science had identified the malaria parasite and explained the role of its vector [carrier], the anopheles mosquito, its conquest was largely a matter of drainage, insecticide, and [window screens]. But drainage in many areas was an expensive project and required a willingness to spend tax funds, and screening presupposed a high enough standard of living for householders to afford screens. In the South it required the infusion of large-scale federal funds under the New Deal and massive outlays during World War II to break the cycle of poverty and relegate malaria [in the United States] to a disease of the past.[23]

The third public health adventure, concerning the metabolic disease known as PKU, was by any measure one of the most dramatic public health victories of the past 50 years. People with PKU (shorthand for phenylketonuria) lack an enzyme needed to metabolize a certain amino acid, phenylalanine, which leads to its buildup in the bloodstream. Too much phenylalanine in childhood can destroy the nervous system, leading to profound mental retardation. In the 1960s, pediatricians learned that if PKU is recognized early enough, children who are fed a special restricted diet—one containing very low amounts of phenylalanine—can grow up normally despite their enzyme defect.

To be sure that all cases of PKU were diagnosed as early as possible, the U.S. Congress ruled in 1966 that all newborns should be routinely screened for PKU. All that is required is a tiny sample of blood, usually taken from a pinprick in the infant's heel; no one need even sign a consent form. This mass screening effort, combined with the PKU diet, has virtually eliminated mental retardation due to PKU—clearly, a dramatic victory for public health.[24]

But the solution of one problem led to the creation of a second one potentially so big that it threatens to eclipse the solution itself. Now, for the first time in history, a generation of young women and men who carry the PKU gene have grown up to become normal adults. Eventually, many of them fall in love and have children. The complication is that, because the PKU diet is so restrictive and expensive (consisting mostly of salads, vegetables, and some costly nutritional supplements), most of these young people have returned to eating normal diets, with their doctors' blessing, around puberty, when the nervous system is fully mature. But the metabolic defect remains. If a woman with PKU gets pregnant while eating a regular diet, her fetus will be exposed to such high levels of maternal phenylalanine that its nervous system will be permanently damaged; it runs a 90 to 97 percent chance of being born retarded. To avoid this enormous risk, PKU women are advised to go back to the restrictive, expensive low-phenylalanine diet, preferably before they even get pregnant, so that embryos are not inadvertently exposed to excess phenylalanine during the vulnerable first weeks of uterine life.

Each of these stories demonstrates the intimate connection of biology, the physical environment, and society. For hookworm a biological disease was completely averted by a social solution. For malaria the physical approach to eradication led to changes in biology—both the parasite and the vector became resistant—leading to what is, at its heart, a lesson in humility. For PKU a social solution to a biological problem carries within it the seed for a future problem. Each story is a vivid demonstration of why the field of public health needs to be in a state of continuous renewal and why its basic approach—measurement, definition of the problem, and design and implementation of a strategy—must be reapplied as conditions change.

Leading educators at schools of public health across the country have struggled for years over just how much training and research ought to be focused on the different perspectives that will be examined in this book—biology, the physical environment, the social environment, the behavioral environment, the health care system, and the global environment. "Should public health identify closely with bacteriology and the successes of the germ theory of disease, or should it seek a broader definition, trying to understand the influence of social, economic, and environmental conditions on the health of individuals?" asks Elizabeth Fee in summarizing this ongoing debate. "Were the social sciences of fundamental importance to understanding the definition, patterns, and distribution of health and disease, or were they a side issue qualifying the serious business of biological research? If public health constituted the study of disease in society, how much attention should be devoted to disease and how much to society?"[25]

These questions are at the core of public health and at the core of this book, which charts its progress over the past 50 years. By recounting some of the most important public health events of this critical half-century, we hope to show the ways in which this ever-changing field has, at least for the time being, acceded to answer them.

Notes

1. The John Snow saga is described in many places, among them George Rosen's *A History of Public Health*, expanded edition (Baltimore: The Johns Hopkins University Press, 1993, pp. 261–263).

2. Harvey Fineberg, personal interview, Oct. 1993.

3. Ibid.

4. "Research and community service at the Harvard School of Public Health," *The New England Journal of Medicine*, 1951, vol. 244, p. 32.

5. William Foege, personal interview, Nov. 1993.

6. Fineberg, personal interview.

7. Samuel Epstein and Beryl Williams, *Miracles from Microbes: The Road to Streptomycin.* New Brunswick, N.J.: Rutgers University Press, 1946, p. 10.

8. William McNeill, *Plagues and Peoples*. Garden City, N.Y.: Anchor Press/Doubleday, 1976, p. 235.

9. Indeed, William McNeill points out that when the Manchu Dynasty collapsed in China in 1911 many Chinese started emigrating to Manchuria, intent on trapping marmots to sell their valuable pelts. But they didn't subscribe to the folklore of the tribesmen, and they trapped sick and healthy rodents alike. The result was an epidemic of plague, which began among Chinese fur trappers and spread along the new railroad connections throughout Manchuria. Luckily, as McNeill writes on p. 156 of *Plagues and Peoples*, medical teams were at work to stay one step ahead of the plague. Without them, "the twentieth century might well have been inaugurated by a series of plagues reaching completely around the earth, with death tolls dwarfing those recorded from the age of Justinian and the fourteenth century, when the Black Death ravaged Europe and much of the rest of the Old World."

10. Judith Walzer Leavitt and Ronald L. Numbers, *Sickness and Health in America*. Madison: University of Wisconsin Press, 1985, p. 9.

11. Edward Shorter, *The Health Century*. New York: Doubleday, 1987, p. 3.

12. Elizabeth Fee, "Designing schools of public health for the United States." In E. Fee and R. H. Acheson, eds., *A History of Education in Public Health: Health That Mocks the Doctors' Rules*. Oxford and New York: Oxford University Press, 1991, p. 158.

13. Ibid., p. 159.

14. The declarations made by the president at the June 1994 commencement were provided by Ned Whalen of the Commencement Office, Harvard University, Cambridge, Mass.

15. Elizabeth Fee, "Public health, past and present: a shared social vision," in Rosen, *A History of Public Health*, p. xxxviii.

16. The early history of the Harvard-MIT School for Health Officers and the Harvard School of Public Health before 1946 can be found in Jean Alonzo Curran, *Founders of the Harvard School of Public Health: With Biographical Notes, 1909–1946* (New York: Josiah Macy, Jr., Foundation, 1970).

17. From an article by Milton J. Rosenau, "Courses and degrees in public health work," *Journal of the American Medical Association*, 1915, vol. 64, pp. 794–796, quoted in Arthur J. Viseltear, "The emergence of pioneering public health education programmes in the United States," in E. Fee and R. M. Acheson, eds. *A History of Education in Public Health: Health That Mocks the Doctors' Rules*, Oxford and New York: Oxford University Press, 1991, p. 134.

18. This period is described by Curran in *Founders of the Harvard School of Public Health*, pp. 19–30, and also by Fee in *A History of Education in Public Health*, pp. 133–142, 163–184.

19. Marie McCormick, personal interview, July 1995.

20. Harvey Fineberg, personal interview, Nov. 1993.

21. Hookworms still infest about 700 million people around the world, especially in poor tropical countries. The disease has been virtually eliminated from the United States.

22. John Duffy, *The Sanitarians: A History of American Public Health*. Urbana: University of Illinois Press, 1990, p. 203.

23. Ibid., pp. 202–203.

24. Richard Koch and Felix de la Cruz, "The danger of birth defects in the children of women with phenylketonuria," *The Journal of NIH Research*, Dec. 1991, vol. 3, p. 61.

25. Fee, "Designing schools of public health," p. 161.

2

Confronting the
Biological Environment

Robert Koch was a slight nearsighted Prussian physician with a scruffy beard and a self-effacing manner. Laboring alone in his makeshift medical-office-cum-laboratory in the 1870s, Koch discovered that a rod-shaped bacillus was the cause of anthrax, a blood disease fatal to sheep, cattle, and humans. His dogged research, as Paul de Kruif noted in the 1926 classic *Microbe Hunters*, proved "that one certain kind of microbe causes one definite kind of disease, that miserably small bacilli may be the assassins of formidable animals."[1] This notion, which has come to be known as the "germ theory of disease," was to revolutionize the fields of medicine and public health.

Koch went on to identify the microbes that caused two of the most deadly diseases of his day: tuberculosis and cholera. His methodical, conscientious work was generally accepted by the scientific hegemony in Germany. But Koch met a formidable foe in Max von Pettenkofer, an elderly physician from München. Pettenkofer challenged Koch to send him some of the comma-shaped germs that he claimed caused cholera, and the old man would demonstrate that they were really harmless. As de Kruif describes the ensuing events:

> Koch sent him a tube that swarmed with wee virulent comma microbes. And so Pettenkofer—to the great alarm of all good microbe hunters—swallowed the entire contents of the tube. There were enough billions of wiggling comma germs in this tube to infect a regiment. Then he growled his scorn through his magnificent beard, and said: "Now let us see if I get cholera!" Mysteriously nothing happened, and the failure of the mad Pettenkofer to come down with cholera remains to this day an enigma, without even the beginning of an explanation.[2]

On this side of the Atlantic, there was a healthy sprinkling of doubting Pettenkofers. In 1884, for instance, an article appeared in the *Journal of the American Medical Association* analyzing bacteriological work on the tuberculosis bacillus, as conducted by Koch and his followers. It concluded that a "too ready acceptance of the bacillus doctrine" could be dangerous, especially since "neither phthisis [an archaic term meaning wasting away] nor any form of tuberculosis is contagious."[3]

From the vantage point of our modern understanding of infection, it is easy to belittle the "anticontagionists" and their earnest objections to the germ theory of disease. But imagine how things must have seemed back in the 1880s. The idea that illness was caused by invisible living things that were passed, by some unseen pathway, from one person to another must have seemed no less bizarre at the time than it would seem today to say that illness comes from creatures from outer space. "Germs" were just as counterintuitive, just as invisible, just as unbelievable as space aliens.

And in the 1880s, there seemed to be little empirical evidence for Koch's theory. Pettenkofer, after all, never did get sick, probably because of some innate resistance to cholera that no one at the time could explain. Indeed, most experience actually seemed to confirm the competing theory of the miasmatics, who believed that illness was caused by putrid air.

But around the turn of the century a natural experiment in Germany allowed contagionists to prove at last that disease sprang from biological pathogens, not harmful vapors. The year was 1892, the setting the old city-state of Hamburg, a self-governing independent entity within the German Reich. Adjacent to Hamburg was the town of Altona, part of the new Germany. Because they were governed by different jurisdictions, the two towns had different water supplies: Hamburg's drinking water came directly from the Elbe River, without being purified in any way, while Altona's water, which also originated from the Elbe, was treated for bacterial contamination at a government-run filtration plant.

In 1892 cholera broke out in Hamburg. The pattern of the epidemic was quite distinctive: along the street that divided Hamburg from Altona, cholera occurred in a nice straight line, affecting the people who lived on the Hamburg side of the street and sparing those on the Altona side. According to the miasmatics, these neighbors, who shared the same air and the same earth, should have been equally at risk of cholera. "A more clear-cut demonstration of the importance of the water supply in defining where the disease struck could not have been devised," notes historian William McNeill. "Doubters were silenced."[4]

Although the Hamburg experience was convincing, it was not scientifically rigorous. To demonstrate that a microbe truly caused a disease could only be done convincingly under laboratory conditions. Eventually, Robert Koch devised a set of criteria—now known as Koch's

Postulates—for proving a cause-and-effect relationship between a particular microbe and a particular disease. Four conditions had to be met: the microbe had to be isolated from every individual with the disease; it had to be grown in culture in the laboratory; then it was to be injected into laboratory animals, in which it had to cause the same set of symptoms as those seen in the original patients; and, finally, it had to be isolated one more time from every infected animal.[5]

As the germ theory of disease was generally accepted among lay people and professionals alike—and as it gained experimental muscle with the application of Koch's Postulates—the age of bacteriology dawned. This revolutionary approach to disease ultimately reshaped our entire concept of what makes people sick and what can be done to keep people healthy. As we will see in this chapter, bacteriology had a distinct effect on public health, separating the old-fashioned sanitarians from the modern scientific scholars. Bacteriology, write historians Elizabeth Fee and Dorothy Porter, "became an ideological marker, sharply differentiating the 'old' public health, mainly the province of untrained amateurs, from the 'new' public health, which belonged to scientifically trained professionals."[6]

In the United States the dawn of bacteriology signified the beginning of what many consider to be the golden age of public health, in which victories against infectious diseases were rapidly ticked off like notches on a gun belt. These victories, along with an occasional failure, form the structure of this chapter, and each of them offers a lesson about larger issues in public health.

- The war on malaria, which created the basic armature on which the current public health hierarchy was built, demonstrated the importance of vectors in the spread of infectious diseases.
- The war on polio, which was fought on two fronts (coincidentally by two groups of scientists at the same school), showed that at the same time public health experts refine methods of treatment (in this case by developing the iron lung), their true Holy Grail is a means of prevention (in this case the polio vaccine) that will render the treatment obsolete.
- The war on smallpox, which mobilized an international team of scientists and politicians in an unprecedented, targeted attack on a microbe, demonstrated for the first (and, to date, the only) time that it was possible to completely extinguish a pathogen through deliberate human activity.
- The war against tropical diseases, which uses techniques as primitive as latrine emptying and as sophisticated as molecular biology, showed how scientific know-how can be put to work to improve the daily lives of people in impoverished developing countries.
- The war on Lyme disease, which emerged seemingly out of nowhere in the late 1970s, revealed the particular hazards of trying to pre-

vent vector-borne diseases, since public health experts must interrupt the life cycles of two or sometimes three nonhuman species to achieve results.

- The war on "river blindness," which is a dramatic story of success and philanthropy, demonstrated the power of a public health strategy combined with a single medication to turn around the fortunes of a despairing community.
- The threat of emerging infectious diseases, which is made most dramatic by the appearance of AIDS, reminds us that humankind's position as the dominant species on the planet is continually undermined by pathogens almost always equipped with more wily survival strategies than ours.

As we review these episodes in the war against biological threats to health, we must begin by remembering the earliest view of the biological environment: as a fetid soup from which a wide range of dank and disturbing vapors—the dread miasmas—arose to pass on strange diseases. The most dramatic refutation of the miasmatics' theories came about as a result of the very disease their philosophy helped to name, because it turned out to be passed around not by vapors at all but by mosquitoes.

Malaria Control and the Public Health Machinery

The miasmatics supplied the name for one of the most deadly diseases known to humans: malaria, Latin for "bad air." Back in Hippocrates' day, malaria was ascribed to the foul air of the marshlands; it was thought to be carried in the vapors that rose up from putrid swamps. Hippocrates believed that malaria, like most other diseases, disrupted the body's four humors: blood, phlegm, black bile, and yellow bile. The specific cause of malaria itself, he wrote, was the drinking of stagnant water.[7]

Hippocrates had it half right. The malarious agent is not in the water itself or the stale air of the swamps. But his observation that regions near swamps or standing water tend to have higher rates of malaria was indeed correct—even though it wasn't until the end of the last century, 2,500 years after Hippocrates, that the full explanation emerged. Malaria occurs near stagnant water because that is where the mosquito that carries the disease is most likely to breed.

Today we know that malaria is a mosquito-borne parasitic infection and that *Anopheles* mosquitoes are the vectors of the malaria parasite. Each time an infected mosquito bites a human it passes on a few of these parasites, known as plasmodia. The plasmodia then lodge in the person's liver and multiply, eventually going into the bloodstream and causing the classical "fever paroxysms" of malaria. These attacks consist of a sudden chill, a high fever with rapid breathing, and finally a sweating stage; after that the fever breaks.

The occurrence of a "paroxysm" means that the plasmodia have left the liver and entered the bloodstream, where they invade and destroy red blood cells. In some forms of malaria each paroxysm involves massive destruction of red blood cells; if the person survives the series of attacks, the malaria is over. But in other forms of the disease the plasmodia hide in the liver for many months and the luckless victim faces a future of recurrent attacks.[8]

Why are certain regions of the world filled with cases of malaria, and others that might have a climate and topography just as hospitable to mosquitoes are malaria-free? It depends, says historian William McNeill, on "local details of the environment," most of them related to the breeding of particular species of mosquitoes:

> Critical variables include the availability of suitable water for hatching the eggs laid by different kinds of mosquitoes. Some species are adapted to spending their larval stage in moving as against still water, and in saline as against fresh. [The] presence and absence of minute trace elements in the water may also play a critical role in determining what sort of mosquito will prevail in a particular locality. In addition, such an unexpected item as the population ratio of human beings to cattle can make a difference. The mosquito species which is Europe's most efficient vector of malaria, for example, prefers to feed on cattle. If enough alternate sources of blood are available to them, these mosquitoes will eschew potential human hosts and thus interrupt the chain of infection, since cattle do not suffer from malaria.[9]

Though malaria is largely a tropical disease, it has been a threat, as we saw in the previous chapter, to Americans in southern regions of the United States, where mosquito populations swell every summer, and in parts of the world where American military personnel are sent to fight. During World War II, both locales were of concern to the U.S. Public Health Service, which is responsible for keeping the military in fighting shape. Many boot camps and other military facilities were located in the South, and many soldiers were being sent to fight in the Pacific, the Mediterranean, and North Africa, regions of the world where malaria was endemic. Malaria was considered a military threat.

Thus began the Malaria Control in War Areas (MCWA) program. Started in 1942 and aimed at draining swamps, a technique already in use during peacetime, and destroying mosquito larvae by any means available, the program was based in Atlanta and within a year was employing 4,300 people, including 300 engineers, sanitarians, and other commissioned officers of the Public Health Service.[10]

"An elaborate anti-malarial doctrine evolved in the war that involved many components that later came to represent anti-malaria programs among civilian populations," says Andrew Spielman of the Harvard School of Public Health. "A manifold structure was established, including scientists of many kinds, particularly entomologists, and medical people who were in the business of prophylaxis [disease prevention]."[11]

Significantly, responsibility for adhering to the structure fell squarely on the shoulders of the commanding officers. According to Spielman, if more than a certain number of troops became infected, their commanding officer would be penalized; he would be subject to disciplinary action if many of his troops were infected.[12]

During World War II, a synthetic drug known as Atabrine was widely used for both prevention and treatment when fighting men were sent to parts of the world where malaria was endemic. Atabrine prevented the development of secondary infections, so even soldiers who had malaria were kept on their feet. But it also had a quirky side effect: it turned people's eyeballs yellow. Consequently, it was easy to spot a serviceman who wasn't taking his medicine.

Toward the end of the war, the fight against malaria took a different tack: killing the vector. This was done with DDT (dichloro-diphe-nyl-trichloroethane), the new pesticide that proved remarkably effective at killing anopheline larvae. DDT was developed in 1940 by Paul Muller, a Swiss chemist who was originally looking for a way to kill clothes moths. Early tests showed that the powder also killed Colorado potato beetles—which at the time were endangering Swiss potato crops—grain weevils, flies, and mosquitoes, with no apparent effect on warm-blooded animals. Noted one malariologist: "It may well be that the two products of greatest importance to emerge from the Swiss Patent Office were Albert Einstein and patent #226,180," the patent for DDT.[13] The world scientific community evidently agreed with this assessment because Muller was awarded a Nobel Prize in 1948.

When Muller's employer, the Swiss pharmaceutical firm Geigy, found that a small amount of DDT contained in a dry powder (about 5% DDT) also killed body lice, the Allied forces began to treat the insecticide as a valuable military secret. Because lice are the vectors of typhus—"the ancient and perennial scourge of unwashed armies in battle and of civilians in war-ravaged cities"[14]—the military greeted the arrival of DDT with great gusto. As Spielman recalls:

> They sprayed it right into the men's clothing. They'd line a bunch of people up on two sides of the room, and people with hand pumps would go down the line pumping DDT powder right up the guy's sleeves. Clouds of dust of the stuff just permeated everything.[15]

By the end of World War II, primarily as a result of the liberal use of DDT and other measures in mosquito abatement programs, malaria was a minor player on the international health scene in many areas, leading to what McNeill called "one of the most dramatic and abrupt health changes ever experienced by humankind." It seemed for awhile that the scourge of malaria might actually be vanquished. But some observers saw the temporary reduction in malaria deaths as a mixed blessing. The resulting growth in population in some impoverished

Spraying with DDT

regions of the world was, as McNeill puts it, "as difficult to live with as malaria itself had been."[16]

The great wartime malaria abatement project helped shape and inspire similar public health measures immediately after the war. In its very structure the mechanism for federal efforts in public health was a direct outgrowth of the Malaria Control in War Areas program. When the war ended, Joseph Mountin and Mark Hollis, supervisor and director, respectively, of the MCWA, convinced their superiors that the program should be converted to peacetime use. In 1946 the U.S. Congress reconfigured MCWA as the Communicable Disease Center (CDC), with Hollis as its first director.[17]

The CDC has kept its acronym intact since then, though in 1970 its formal name changed to the Center for Disease Control, in 1980 to the Centers (plural) for Disease Control, and in 1992 to the Centers for Disease Control and Prevention.

A military spirit pervaded the public health effort in more subtle ways as well. Spielman recalls that after the war a huge stockpile of unused DDT was housed at Otis Air Force Base on Cape Cod. The temptation to use the pesticide was too hard to resist.

> They had all these bombers that were no longer dumping explosives on Germany, so they used them to dump DDT on the Cape. The war mentality, which carried over, led to profligate use of DDT aimed at wiping out all mosquitoes.[18]

23

Buoyed by its earlier success in controlling malaria, the World Health Organization, with support from the U.S. Congress, mounted an international program in 1957 to eradicate malaria throughout the world. The program was instituted as a five-year operation to be coordinated by the Agency for International Development with the cooperation of the Pan American Health Organization, the World Health Organization, and UNICEF. It was instituted along the lines proposed earlier by Paul Russell, then of the Rockefeller Foundation, whose impassioned document written for the International Development Advisory Board had laid out the rationale for an immediate and concentrated global effort at total eradication. According to Russell, timing was crucial:

> This is a completely unique moment in the history of man's attack on one of his oldest and most powerful disease enemies. Failure to proceed energetically might postpone malaria indefinitely. . . . Eradication can be pushed through in a community in a period of eight to ten years, with not more than four to six years of actual spraying, without much danger of resistance. But if countries, due to lack of funds, have to proceed slowly, resistance is almost certain to appear and eradication will become economically impossible."[19]

The five years authorized for the international program, however, turned out to be insufficient. In 1963, when American funding for the program was scheduled to end, malaria was reduced to its lowest rate ever throughout the world, but it was not low enough. As John M. Karefa-Smart, an alumnus of the Harvard School of Public Health and head of the Sierra Leone delegation to the United Nations, reported that year, malaria was still the most significant health threat on the African continent,

> not only because it takes its toll of life from the first week of infancy till the last days of old age, but also because most of its victims whom it does not kill are left more or less handicapped for life—having achieved an uneasy truce with the malaria parasite within the body— with depleted energies and with malfunction of nearly every organ of the body.[20]

One of the reasons that malaria was never truly vanquished is that mosquitoes evolved DDT resistance. Pesticides developed subsequently were effective against DDT-resistant mosquitoes, but they never quite achieved the vanquishing power of DDT. At the same time, plasmodia became resistant to the quinine derivatives used to treat and prevent malaria, and scientists scrambled to synthesize chemical formulations that retained the protective properties of Atabrine and its relatives while containing enough differences to kill off resistant plasmodium strains.

While the military approach of total surrender seemed appropriate at the time, total eradication of malaria proved far more elusive than total eradication of smallpox (which, as we will soon see, was actually accomplished in 1977) or polio (which is targeted for the year 2000).

Indeed, by the 1980s, malaria had again become the most common and lethal infection in the world, affecting 300 million to 500 million people each year and killing at least 3 million, most of them children.[21] As Spielman explains:

> The anopheline vector tends progressively to resist insecticides, the pathogen loses susceptibility to drugs, financial appropriations become exhausted, herd immunity [a generalized collateral protection against disease] declines, the operational staff becomes demoralized, and the subject population loses interest in the effort while learning to expect a relatively disease-free state.[22]

Obviously, nature does not stand still just because scientists have come up with a stop-gap control measure for an infectious disease. The parasite, the vector, and the social structure all continue to evolve. The only known way to truly control an infectious disease is to develop an effective vaccine. This is the underpinning of the classic approach to disease prevention—prevention through inoculation. And the search for an effective vaccine had one of its most dramatic incarnations in the 1950s, when the nation was transfixed by the race to develop a vaccine against polio.

Conquering Polio: From Iron Lung to Cell Culture to Vaccine

In the darkest moments of the age of polio, whole communities were gripped by fear. The casual joys of summertime—swimming in the town pool, going to an overnight camp, sitting in a cool movie theater for an afternoon double feature—became times of foreboding. Wherever children gathered in the 1920s through the early 1950s, especially during the summer months when the polio virus was in season, the dreaded paralytic disease could spread. Worried parents kept their children home, summer after summer, and watched for signs of illness.

Polio was an unusual infectious disease because it actually grew worse as overall health conditions improved. In overcrowded tenements and rural slums, young children were exposed to the poliovirus within the first few years of life, usually as the virus was shed from carriers through their feces. At these early ages of exposure, polio caused virtually no symptoms beyond a transient fever, much like a case of the flu.

Children of the middle and upper classes were protected from exposure to the virus during their infancies. In the first half of the century, before day care, most very young children of means stayed at home with their mothers or governesses and rarely interacted with large groups of peers. When they were older and went to school or summer camp or swam in public pools, they encountered large groups of children for the first time.

In countries or even urban American slums with poor sanitation and squalid living conditions, writes historian Harry Dowling, polioviruses were widely disseminated, the disease remained endemic, and virtually all children had acquired infection and immunity by the age of five years. The price for this immunity was a few cases of true infantile paralysis, but by far the majority of infections were inapparent. In such a population no epidemics occurred; indeed, poliomyelitis was hardly recognized.[23]

But first-time exposure to the poliovirus at an older age was more likely to lead to some of the most pernicious complications of infection: muscle stiffness, pain, and eventual paralysis. This was the way the disease was showing up in the most advanced countries in the world, such as Sweden and the United States. And this is why, to the middle and upper classes who helped mold public opinion, polio was, in the middle of the twentieth century, an especially dreaded word.

The idea that this ruthless killer was stalking the nation's children, especially its privileged children, was more than most Americans could bear. As a result,

> quack remedies abounded. Frog blood; radium water; wine of pepsin; a drink made of rum, brandy, champagne, and mustard plaster (which presumably either nauseated or inebriated patients to the extent that they didn't care if they caught the disease)—all these were offered as cures. A former state legislator claimed that cedar shavings scattered about a house would protect children and sold bags at a profit. He was fined $250 and sentenced to thirty days in jail. When fresh ox blood was touted as a cure, people showed up at East Side slaughterhouses with buckets ready. Nothing worked—nothing, that is, except the coming of fall.[24]

The images of those polio summers still haunt many who lived through them—empty swings in suburban playgrounds; "CLOSED" signs at city pools; and, most dreadful of all, hospital wards filled with row upon row of mammoth artificial breathing machines, a sweet and frightened young face peering out of each one.

Those machines were the legacy of one brave eight-year-old girl who was dying of polio in a Boston hospital in October 1928. Her chest muscles were paralyzed and she was suffocating, unable to get her lungs to expand enough to take in air. She had the panicked look of a trapped animal; her fingers and face were blue.

In her room at Children's Hospital sat a huge tin box with a hole at one end and a motor at the other, whining and humming incessantly. The staff had moved the machine into her room because they knew she would need it soon and wanted her to get used to its noises. By Sunday morning, when the girl slipped into a coma from lack of oxygen to the brain, she was already lying inside the machine. But the staff was uncertain about what to do next and afraid to turn the thing on. So they telephoned the machine's inventor, Philip Drinker.

Drinker, an engineer and a professor of industrial hygiene at the

Harvard School of Public Health, was at the hospital in minutes. He had invented the big tin box as an artificial breathing machine, and this was the first time it would be tried on anyone except himself. It was officially called the Drinker Respirator, but most Americans came to know it by its more descriptive name—the iron lung.

"Phil started the pump, and in less than a minute saw the child regain consciousness," wrote Drinker's sister, the biographer Catherine Drinker Bowen, in an affectionate memoir called *Family Portrait.* "She asked for ice cream. Phil said he stood there and cried."[25]

This little girl was a pioneer, the first polio patient to try the iron lung. Sadly, she subsequently died. But the next person to use the machine lived—and thrived for at least another 30 years, at which point Drinker lost track of him. That second polio patient was Barret Hoyt, a senior at Harvard, who was put into the respirator when he was on the verge of suffocation. After a few minutes inside the iron lung, Hoyt's first grateful words were "I breathe."

Getting Philip Drinker to the School of Public Health had been something of an iconoclastic move. His brother, Cecil, was a professor of applied physiology and assistant dean at the school. Ten years earlier, while he was acting chairman of the Department of Physiology, Cecil had decided that the school needed an engineer on its faculty and that Philip would be ideal for the job. But Phil had no particular expertise in health or medicine. As Catherine Drinker Bowen put it, Harvard research physiologists at the time "regarded the engineering sciences

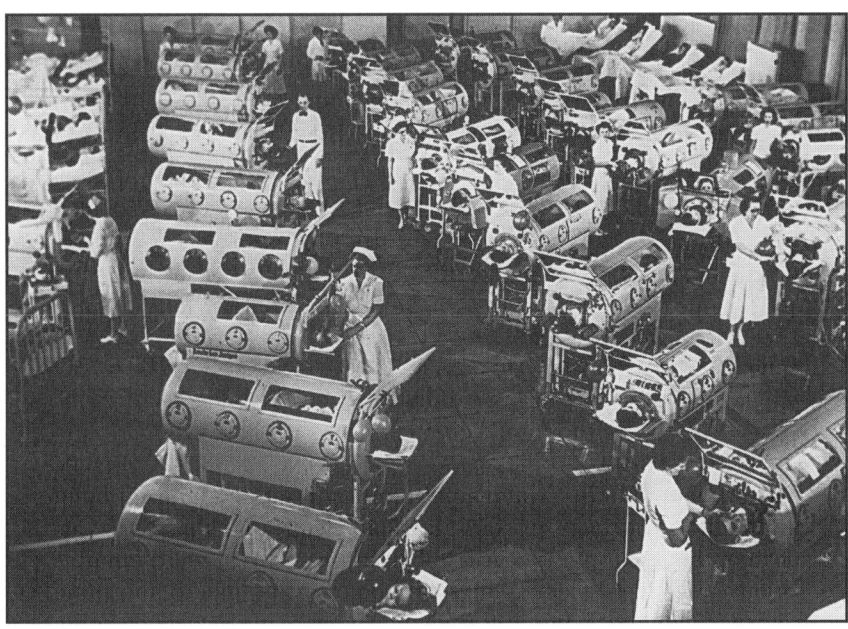

Philip Drinker's iron lung

somewhat as philosophers of the sixteenth century looked on barber-surgeons—useful creatures but hardly to be included in the guild of scholars."[26]

By nominating his brother to join the faculty, Cecil Drinker took a big risk but at the same time effectively broadened the definition of the mission of public health and the disciplines that should be brought to bear on that mission.

The idea for the Drinker Respirator arose, as do so many great discoveries, almost by accident.[27] One morning in the fall of 1926, Drinker was watching a colleague, Louis A. Shaw, work on a device to measure breathing in an animal. The device was a small metal box, about the size of a shoe box, with a hole on one end and a water tube, called a manometer, on one side. Shaw placed a cat inside the box with its head coming out of the hole; a snug rubber collar fit around the cat's neck to keep the box airtight. The manometer measured the strength of each breath by the cat by the rise and fall of the water level. When the cat exhaled, the water level rose; when the cat inhaled, it fell.

Drinker watched Shaw at work and thought of a use for the box that went beyond measurement. What if the cat *could not* breathe? Could pressure be applied inside the airtight box to force the cat's chest to rise and fall, thus stimulating breathing artificially? After he and Shaw jury-rigged a way to pump air into and out of the box, they found that cats who had been paralyzed with an injection of curare could be kept alive, the machine breathing for them, for hours at a time.

Drinker secured $500 from a New York gas company to see if his contraption could be used in cases of asphyxiation caused by gas leaks or suicide attempts. He hired a Boston tinsmith to make a box to fit a man—to fit Drinker himself, as it turned out, since he would be the first guinea pig. Drinker cobbled together an air pump for the box by rooting through a closet at school filled with old vacuum cleaners that had been discarded by a fan manufacturer. Two of the vacuum cleaner motors, plus plenty of tape, made a pump for the machine; a hand-operated valve could then be used to force air in and out.

All that was left was to make a rubber collar—one tight enough to maintain the seal but loose enough not to cut off blood flow to the brain. This problem was a stumbling block for many months. It was not until 1928 that Drinker was ready to slide into his box, rolling in on a mechanic's dolly borrowed from a local garage, his head resting on two seat cushions on the floor, a rubber collar coiled tightly around his neck. With the pump noisily whirring away, Drinker took pure oxygen into his lungs for 15 minutes, with Louis Shaw at the machine's controls. When the experiment was over, Drinker's chest did not move visibly for seven minutes, until he had enough carbon dioxide in his oxygen-rich blood to need to take a breath. "At that moment," he wrote in his report, "I resumed regular breathing and we terminated the experiment."[28]

Once news got out of his respirator's success with polio victims, Drinker became a hero overnight. His machine was written up in newspapers from New York to California, from the *Peiping and Tienstsin Times* in China to *L'Illustration* in Paris. "At the Chicago World's Fair the iron lung drew crowds to the Hall of Science," Drinker's sister recalled. "In his daily comic strip, Dick Tracy rushed a child to the machine."[29]

So Philip Drinker, a quiet, modest man, became an international celebrity, hailed as the savior of thousands. In retrospect, he thought the fuss excessive. He told his sister years later that one of the highlights of his career came in 1953, when he shared a platform with Jonas Salk. He thought the true honor belonged to the developer of the polio vaccine, not to the inventor of the iron lung. The iron lung "couldn't prevent the cases," he said. "It could only treat them."[30]

Others, of course, also saw the value of prevention rather than treatment, and scientists across the country were frantically working on a vaccine in the 1930s and 1940s. As anyone could see, polio was the single most emotional topic in all of biomedical research and whoever discovered a vaccine would be as revered as the person we imagine today would be who finds the cure for AIDS—an instant and permanent folk hero. That was heady stuff for scientific researchers, especially combined with the fact that money was being made available to work on a polio vaccine, through the National Institutes of Health, several pharmaceutical companies, and the National Foundation for Infantile Paralysis, the private philanthropy known to millions of small donors as the March of Dimes.

But until the mid-1940s, research on a polio vaccine was stalled due to one critical problem: no one could grow the poliovirus in a form that would permit mass production of vaccine. Albert B. Sabin, a young researcher at the Rockefeller Institute, had succeeded in growing the virus in human nerve tissue, but that was impractical for two reasons: nerve tissue was difficult to come by and, more importantly, any vaccine manufactured from such a tissue culture would contain remnants of foreign nerve cell proteins. If nerve cell proteins were injected along with the vaccine—as they almost inevitably would be—they could lead to an allergic and potentially fatal brain inflammation that was more dangerous than polio itself.[31]

Obviously, if a polio vaccine were to be developed, the first step was to find a better way to grow the poliovirus in the laboratory. That is where Thomas Weller comes in. Weller, a recent graduate of the Harvard Medical School, was in 1947 in charge of the day-to-day operations of the virology laboratory, headed by the noted John Enders at the Harvard-affiliated Children's Hospital. "I was interested in chicken pox virus," recalls Weller, who by 1954 was chair of the Department of Tropical Public Health at the Harvard School of Public Health, a post he held until his retirement in 1981. "I was trying to grow the virus in tissue culture, which no one had done before."[32]

Weller had refined a variation of a culture technique that involved suspending human tissue—obtained from miscarried fetuses or still-births from the Boston Lying-in Hospital—in a culture medium. His first success occurred with mumps virus, as a result of an offhand comment by Enders. As Weller recalls:

> At that time viruses had been grown in tissue culture, often by the Maitland technique, which was a little flask with suspended fragments of tissue in it. They subcultured [discarded] the tissue every one or two days. I got to thinking about that, and I thought if you were working with a slow-growing virus, you might be discarding it every time you threw away the tissue. I thought, "Maybe we should leave the tissue fragments in there and change the fluids instead." I brought up my idea with Dr. Enders and he was intrigued. He said, "Before you do that, why don't you try growing mumps virus in tissue culture?"[33]

Only after he succeeded did the young Weller learn that he was the first scientist to culture mumps virus in the laboratory. He was helped in large part by the availability of antibiotics, whose manufacture had been perfected during World War II. Applying them liberally to his tissue cultures, he was able to keep them free of bacterial contamination for long periods of time.

Now he was ready to try transferring his success with mumps virus to chicken pox virus, formally known as varicella. One day in 1948, after he had he inoculated several test tubes with varicella, Weller saw that he still had four flasks that contained human embryonic tissue suspended in a nutrient broth. Into those four flasks he introduced a strain of polio virus that had been preserved in the laboratory freezer. The varicella cultures never took, but, remarkably, the polio cultures did. As Weller says, with characteristic understatement: "To our surprise the virus grew and paralyzed mice, showing it was growing and could be serially passed."[34]

Weller's discovery unleashed a flurry of activity in polio vaccine research, which until that moment had been hampered by the need to grow virus in live animals (and research monkeys at the time cost $75 each) or brain cells. In 1954 the three scientists responsible for this revolutionary discovery—John Enders, Tom Weller, and Frederick Robbins, Weller's colleague and roommate from medical school days—shared a Nobel Prize. The award cited them:

> for their discovery of the ability of poliomyelitis virus to grow in cultures of various types of tissue. . . . These discoveries incited a restless activity in the virus laboratories the world over. The tissue culture technique was rapidly made one of the standard methods of medical virus research.[35]

The Nobel Prize came just months after Weller was named the Richard Strong Professor of Tropical Public Health at Harvard, as well as

Thomas Weller accepting the Nobel Prize

chair of the department. One of his colleagues at the time, Eli Chernin, remembered the day of Weller's award vividly:

> Late in that autumn day in 1954 when he learned of the Nobel Prize, a subdued 38-year-old Weller returned to his new Department after an exhausting round of meetings and interviews at the nearby Children's Hospital where the prize-winning research had been done. As his colleagues gathered around—none of us had seen a Nobelist before—Weller slumped into a chair and remarked to no one in particular, in his quiet uninflected voice, "Now, I guess we'll have to show them it wasn't a flash in the pan."[36]

It certainly wasn't. Subsequent discoveries in the department were legion. Investigators isolated and grew the rubella virus, providing the basis for eventual vaccines, and isolated the human cytomegaloviruses, linking them to causes of congenital brain damage. And Weller succeeded in his original goal—to isolate and grow the chicken pox virus.

Meanwhile, researchers elsewhere were using the Enders-Weller-Robbins tissue culture technique in a mad rush to find a polio vaccine. In the early 1950s scientists at Johns Hopkins University, Yale, and the Universities of Toronto, Pittsburgh, and Cincinnati[37] were engaged in fertile collaborations—peppered with a healthy dose of competition—to develop a vaccine based on either a killed polio virus or an attenuated (weakened) one. The killed-virus vaccine was made up of poliovirus that had been treated with formaldehyde so that it lost all its infectious properties. The virus did retain its shape, though, which stimulated the body to make polio antibodies. The live-virus vaccine worked much the way the smallpox vaccine worked, transmitting a harmless case of the disease to stimulate antibodies that would successfully fight off a more potent infection in the future.

In the spring of 1954 the killed-virus vaccine was considered ready for widescale testing. Several cautious scientists thought such a test was premature. They argued that the vaccine had not been adequately tested for safety. How could anyone be sure that the killed strain was safe? For that matter, how could anyone be sure that a killed virus would protect vaccinated children when they encountered the real thing?

These questions were quickly pushed into the shadows. Public pressure to move on the polio vaccine was intense.[38] The articulate, politically mobilized, influential middle class saw another polio season looming and was not about to risk its young children's lives while a potential cure languished in the laboratory, slowly wending its way through bureaucratic hurdles on the way to the marketplace.

Bowing to the political realities, the National Foundation for Infantile Paralysis funded a large-scale study of the killed-virus vaccine developed by Jonas Salk at the University of Pittsburgh. The trial, coordinated by Salk's mentor, Thomas Francis at the University of Michigan, involved a placebo-control trial of 420,000 children in 11 states and an observational study of 230,000 children in 33 states. The first, second, and third graders who received all three Salk vaccine injections received buttons proclaiming them polio pioneers.[39]

One year later, on April 12, 1955, the foundation called a press conference in Ann Arbor, Michigan, to announce the results. It was the tenth anniversary of the death of Franklin D. Roosevelt, the nation's most famous polio victim. In what was described as "an atmosphere of scurrying reporters and popping flash bulbs,"[40] Francis announced that the Salk vaccine had proved to be 60 to 90 percent effective against paralytic polio. Later that day in Washington, D.C., Oveta Culp Hobby, secretary of the newly created U.S. Department of Health, Education, and Welfare, made an announcement of her own. Hobby, an old friend of President Dwight D. Eisenhower and the only woman in his cabinet (even though he was a Republican and she was a lifelong Democrat), gave approval for a national immunization campaign, granting licenses to six pharmaceutical companies to begin manufacturing and distributing killed polio virus vaccine according to Salk's instructions.[41]

Though the initial response was laudatory, Hobby quickly came to regret that action. By the end of April the government began receiving reports of trouble with the new polio vaccine. Some vaccinated children were coming down with active cases of paralytic polio. These were not merely cases of children with inadequate protection being exposed to polio in the community; these were cases of children contracting polio directly from the supposedly safe vaccine. As the tragic reports piled up, what had begun as a triumph of modern public health quickly turned into a national nightmare.

On May 7, 1955, Hobby's surgeon general, Leonard Scheele, called for a halt to distribution of the vaccine. Government scientists scrambled to piece together what was going wrong. The statistics were devastating: 79 children and 105 of their close contacts had come down with polio. Of these cases, 141 people became paralyzed and at least 11 died.[42] What all had in common was that they had received their vaccines from a batch produced by Cutter Laboratories of Berkeley, California. Something in the Cutter method of inactivating the polio virus had failed, and the children had been injected, not with "killed" virus, but with infectious polio.

Soon after the incident, Cutter Laboratories was temporarily closed and Secretary Hobby resigned. Somehow, Surgeon General Scheele (who resigned the following year) managed to keep parents from panicking and to convince the American public that the Cutter contamination was an aberration, unlikely to happen again. Maybe because they were so afraid of polio, Americans were willing to believe that the vaccine, no matter how rocky its beginning, would ultimately save their children. Within a few months, vaccine production resumed and soon the Salk vaccine was coming into general use.

Historian Harry F. Dowling summarizes the remarkable flurry of activity over polio in the 1940s and 1950s as follows:

> The conquest of poliomyelitis had all the elements of an American success story: the focus on an objective so simple in concept and so obviously beneficial to mankind as to resemble the quest for the holy grail; the enlistment of all groups and classes in the crusade; the use of the mass media for publicity; the appeal to sentiment; the thorough organization of the campaign; the unwillingness to leave the final decisions on research support in the hands of the experts; and the popularization of a hero as David, the killer of the giant.[43]

But the dominance of the Salk vaccine did not last for long. Just around the corner was a different polio vaccine, this one given orally instead of by injection and using an attenuated strain of live polio virus. The advantages of this new vaccine, developed by Albert Sabin at the University of Cincinnati, were twofold: its ease of administration and its hidden benefit to nonvaccinated members of the community. Because Sabin's vaccine used a live virus, active virus particles were released into the community through the excretions of vaccinated individuals. This exposure produced a collateral protection even in the unvaccinated, in numbers great enough to reach a critical mass of immune individuals, known as "herd immunity," which keeps overall disease rates down. As medical writer Greer Williams wrote in 1959, when the Sabin vaccine was still under development:

Confronting the Biological Environment

Sabin frequently has pointed out that the Salk vaccine does not eradicate the ubiquitous polio virus from the bowels of the vaccinated population. The Salk vaccine builds antibodies in the blood, and these head off paralysis, but the virus apparently can still live and multiply in the intestinal tract. In contrast, Sabin claims that a live-virus vaccine given by mouth—eliminating the needle—will keep wild polio viruses from passing through by creating a local, or cellular, resistance in the walls of the intestines. Thus, he assumes that a live-virus vaccine ultimately could lead to the extinction of paralytic polio viruses by eliminating any possible carrier state.[44]

The Sabin live-virus vaccine was introduced in the early 1960s and eventually eclipsed the Salk vaccine. Although infection by the weakened vaccine virus accounts for a handful of polio cases in the United States each year, scientists generally prefer the live-virus vaccine because of the more natural immunity it confers to the vast majority of people who receive it. With the Sabin vaccine, polio is on its way to being eliminated altogether from the Western hemisphere. When that happens, it will join a highly select club of viruses that have been vanquished by vaccines. Right now, that club has one single member: the smallpox virus.

The End of Smallpox

Eradication of smallpox required more than a good vaccine. It required a sophisticated strategy for distributing the vaccine, an international collaboration to get it to the places that most needed it, and the good fortune of squaring off against a relatively easy microbial foe: a virus with no natural hosts other than humans. All these factors finally put an end to smallpox. In October 1977 a hospital cook in Somalia named Ali Maow Maalin contracted smallpox from two Ethiopian children who had fled to Somalian refugee camps after their nation's civil war. Maalin, who recovered, turned out to be the last natural case of smallpox on earth.

Getting to that point was not easy. At the close of World War II, smallpox was still endemic in at least 60 countries: parts of the United States and Mexico, England, France, Germany, Spain, Portugal, Italy, Japan, Indonesia, the Middle East, and almost every country in South America and Africa.[45] In the subsequent years, however, two major technological innovations changed the face of smallpox immunization forever. In 1949 freeze-dried vaccine was introduced, which made it possible to transport smallpox vaccine to remote rural areas without refrigeration. And in the early 1960s the jet injector was developed and adapted for smallpox, allowing for painless injection of vaccine so swift that a single health care worker could vaccinate up to 1,000 people an hour.[46]

By the mid-1960s, smallpox was concentrated in only five major zones worldwide. North America, Central America, and China were free of the disease. It seemed possible to eradicate smallpox internationally, which had been the goal set by the World Health Organization back in 1958. In 1966 the WHO established an international coordinating team to get serious about carrying out this goal.

In charge of the team was D. A. Henderson, who later became dean of the Johns Hopkins University School of Hygiene and Public Health and subsequently science adviser to President George Bush. Second in command was William Foege, who had recently

Donald Hopkins

received a master's degree in public health from Harvard and would, under President Jimmy Carter, join a long line of Centers for Disease Control directors—five in a row, whose combined tenures spanned more than a quarter of a century—who were graduates of the Harvard School of Public Health.[47]

Donald Hopkins (who later became part of this procession, first graduating from the Harvard School of Public Health and later becoming a member of its faculty and serving as acting director of the CDC), led the smallpox eradication program in Sierra Leone, at the time the country with the highest smallpox attack rate in the world. In 1967 fewer than half the people of Sierra Leone had smallpox vaccination scars, and less than 25 percent had been vaccinated in the previous five years. At first, Hopkins and his colleagues tried to vaccinate everyone using the jet injectors, gathering as many people as possible in central villages. Their goal was to vaccinate 80 percent of the population. After about a year they could see that their efforts weren't working.

At this point, Hopkins, his colleagues, and their supervisors, Henderson and Foege, came up with an alternative strategy. They called it "selective epidemiologic control"—concentrating their vaccination efforts on households and villages where smallpox outbreaks were already under way. As Hopkins recalls:

> This new approach . . . allowed us to focus on about 5 percent of the population, including people with smallpox and their immediate household contacts, rather than the entire population. This new strat-

egy was put into effect in Sierra Leone in August, 1968, and smallpox was eliminated from Sierra Leone nine months later, only 15 months after the beginning of the program in that country. Similar rapid progress occurred in several other West African countries. By May, 1970, the incidence of smallpox in the 20 countries in West and Central Africa to which the U.S. was providing assistance had plummeted to zero.[48]

In the end, the final two countries, Ethiopia and India, presented the greatest obstacles. The last country in Africa to have smallpox, Ethiopia had some of the most rugged mountain terrain on the continent, where isolated villagers lived in fear and suspicion of strangers. Volunteers from the United States, Japan, and Austria joined Ethiopian sanitarians on heroic forays into the mountains for stretches of 30 days at a time, carrying their personal and medical supplies on donkeys or on their own backs, dusting their bodies liberally with DDT powder or wearing dog collars to keep lice and fleas away.[49]

In India selective epidemiological control ran headfirst into theology. The Hindu goddess of smallpox, Sitala, often took priority over the scientific approach of vaccination. Still, the public health scientists managed to outwit the gods. They instituted a 24-hour surveillance on every household and offered a 100-rupee reward (equivalent to about 20 percent of the average annual income) for anyone reporting a case of smallpox to the authorities. These efforts led to a swift and dramatic decline in the disease. In May 1974 there were almost 10,000 active outbreaks of smallpox in India. By mid-March 1975, there were only 63.[50]

When he returned to the United States, a fired-up Hopkins enrolled for graduate studies at the Harvard School of Public Health. He received a master's degree in 1970 and taught in the school's Department of Tropical Public Health until 1977. The global eradication of smallpox, decisive as it was in shaping the career goals of Hopkins and perhaps others involved in the effort, was a unique opportunity for public health that is unlikely ever to be repeated. The epidemiology of smallpox "lent itself to eradication," notes Hopkins:

> There was no animal reservoir from which the disease could be reintroduced if it were eradicated from the human population. Essentially all cases of smallpox, particularly those who were infectious, were readily detectable. No expensive tests were necessary to decide whether or not a person had smallpox. The two-week incubation period allowed tracing of cases, getting to contacts, and immunizing them to prevent infection. Patients were only infectious at most for a three-week period, so it was not a lingering type of infection. Moreover, the disease conferred natural immunity to patients in communities that were so unlucky as to get the disease.[51]

Once smallpox had been eradicated worldwide and the final vaccination given (in 1982), scientists were faced with an unprecedented quandary: What should they do with the smallpox virus? The only repository for the virus was in two high-security laboratories, one at the CDC in Atlanta and the other at the Research Institute for Viral Preparations in Moscow. Stockpiles of smallpox had been restricted after two tragic laboratory accidents in England in the 1970s that spread the lethal disease to five unfortunate souls. In March 1973 a medical technician contracted smallpox from the virus he was handling; he passed the disease on to two contacts, both of whom died. In August 1978 a female photographer taking pictures in a medical laboratory contracted smallpox through a ventilation duct linked to the lab of prominent scientist Sir Henry Bedson. The photographer died; her mother also developed smallpox, but she recovered. Sir Henry, overcome with remorse over the young woman's death, later committed suicide.[52]

The last reserves of smallpox were slated to be destroyed on the last stroke of midnight in 1993. "We cannot really say we have rooted out this disease from the planet," said Russian virologist Yuri Ghendon, "until we have done away with its cause."[53] But at the eleventh hour, the smallpox virus was given a reprieve, and the world scientific community decided to debate some more the pros and cons of deliberately wiping out a biological species. Those who wanted to keep the virus alive said there was still much to learn about variola's virulence, pathogenesis, and potential for screening drugs for other viral diseases. Those who wanted to execute it said most of that could be learned by using cloned variola DNA or the nucleotide sequences that had already been worked out for several sections of the virus. (Of course, as Ghendon pointed out, "Why anybody would want to study the virus of a disease that no longer exists, when there are microorganisms emerging or re-emerging that are real public health threats today, beats me."[54])

In October 1994 a committee of the World Health Organization voted on the fate of the world's smallpox samples, which at the time consisted of 700 small plastic vials in liquid nitrogen freezers, 450 vials in Atlanta and 250 in Moscow. The committee recommended that the vials be autoclaved and incinerated (that is, boiled and burned) on Friday, June 30, 1995.[55] A few months later, in January 1995, the variola virus got another reprieve. The executive board of the World Health Organization decided to delay any decision about smallpox. In the meantime, as Yuri Ghendon put it, the safety of humankind was at risk. Now that humanity has lost either natural or artificial immunity to variola, he said, the accidental or deliberate release of the smallpox virus could be absolutely devastating. "If there was an outbreak today," Ghendon told a reporter for *Science* magazine, "about a million people could die or go blind from the disease in the several months it would take to get enough vaccine produced."[56]

As we have seen, smallpox held on for the longest time in the developing world, where distribution of vaccine was difficult and political and cultural mores occasionally interfered with the identification and quarantine of new cases. The less developed regions of the world still bear an unfair burden of infectious diseases, and many of them have no effective vaccines in the offing for the illnesses that most plague them. For these infections, the available methods of control are more complex: either control the vector, develop treatments that limit the virulence of the disease, or create a vaccine that is not only effective but also easy to administer, heat resistant, and cheap.

Out of the Tropics: Schistosomiasis and Leishmaniasis

The developing nations of Asia, Africa, and South America have at least two strikes against them in terms of public health. For one, their climates are tropical, which breeds some of the most deadly vector-borne diseases on earth. Two, their poverty leads to living conditions in which these vectors and the diseases they carry are most easily spread. In terms of sanitation, sewage treatment, water purity, and housing stock, the countries of the less developed world are akin to Europe and North America in the 1700s.

One of the diseases endemic to regions that are both hot and poor is schistosomiasis. The disease is caused by a worm-like parasite called a schistosome, or blood fluke. Its eggs are deposited into ponds and streams and hatch to release free-swimming larvae that penetrate and develop inside tiny freshwater snails. A subsequent larval stage, which matures in the snails, is capable of penetrating human skin. The larvae then invade the skin of someone standing or bathing in infested waters and migrate to the person's intestines, where they mature into adult worms that mate, develop eggs, and begin the cycle anew.[57]

"If the male and female end up in your body they mate, copulating constantly, producing and shedding eggs 24 hours a day," says John David of the Harvard School of Public Health. "They are locked in embrace forever. When you defecate in the water, the eggs pass out and hatch, and they swim around and enter into tiny snails."[58]

Schistosomiasis is almost as common as malaria. Estimates are that 200 million people carry the schistosome worm and that 800,000 people a year die from the infection.[59] The worm does its damage indirectly. When female worms deposit their eggs in a person's tissue, the body re-

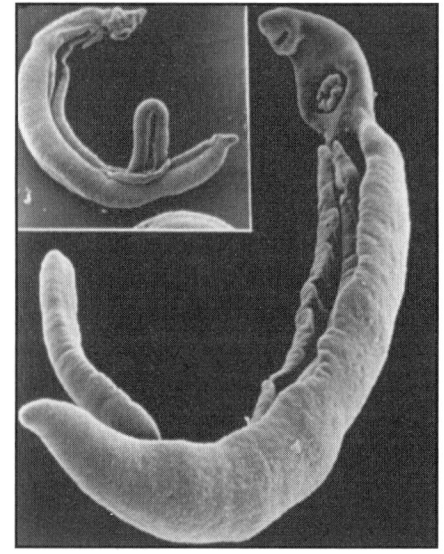

Schistosome parasite

John David

sponds by walling off the eggs into cysts known as granulomas. The granulomas are what lead to problems: blocking capillaries, impeding blood flow, causing scarring. Too much scar tissue in internal organs can result in liver cirrhosis, bladder tumors, kidney failure, and death.[60]

Prevention of schistosomiasis should be easy. As David puts it, "Theoretically it's a totally social thing"; all that it should take to stop the cycle of schistosomiasis is for people to have and use toilets. If they are not defecating into the water they wade in, transmission of the worm is short circuited.

But in the absence of the costly approach of massive waste disposal systems, a drug now exists that can be given just once a year to keep the worms at bay. "There's always the question of creating drug-resistant parasites," says David, pointing out that some have already been found in Africa. But for now the drug, known as PZQ (praziquantel), has significantly reduced the mortality and morbidity rates for schisto-somiasis in Brazil, the Philippines, Sudan, and China.

Even with PZQ, though, the scourge of schistosomiasis is far from over. Most patients become reinfected within one year. The real hope for eliminating the disease is the one that has worked so well with infec-tious diseases in the past: finding a new vaccine. One possibility, the Sm23 DNA vaccine developed by Donald Harn of the Harvard School of Public Health, uses the DNA from the schistosome itself that codes for one of its surface proteins. This DNA vaccine has been shown in mice to stimulate antibodies to the schistosome, without the hazards of inadvertent infection posed by either killed-virus or live-virus vaccines.[61]

39

Another tropical disease, leishmaniasis, has not been as responsive to public health attack. Leishmaniasis is caused by a protozoan (a single-cell parasite) carried in the sand fly, which is plentiful in India, China, Sudan, Ethiopia, Iraq, the Mediterranean, and Brazil.[62] When the sand fly bites a person, it transmits the leishmania protozoa. In the sand fly saliva are substances, too, that allow the leishmaniasis infection to take hold: chemicals that interfere with white blood cell activity in the host and a hormone-like molecule that dilates the host's blood vessels.

In South America, leishmaniasis comes in several forms: one that causes ulcers on the skin, from which most people eventually recover, and another, known as mucocutaneous leishmaniasis, that is enormously destructive. The mucocutaneous form, according to David, who has studied it in Brazil, "can metastasize and completely destroy the nose, the larynx, the palate. It's a very destructive, disfiguring disease." A third form is visceral leishmaniasis. In the Mediterranean and North Africa, leishmaniasis takes a different form; it is known there as kala azar, and it attacks the internal organs. Kala azar frequently is fatal.

Public health officials have a small bit of ammunition against leishmaniasis, an old family of drugs known as pentavalent antimonials. Made of heavy metals, the drugs can lead to nausea, vomiting, and other serious side effects. But someone who knows at the start that his or her leishmania was caused by the dangerous, disfiguring species might well decide that some vomiting now is worth preservation of facial structures later. To allow people to make these informed decisions requires early identification of the species of the parasite causing a particular infection.

In 1983 molecular biologist Dyann Wirth of the Harvard School of Public Health spent some time in the Brazilian state of Bahia, where several forms of leishmaniasis were endemic. At the time, diagnosis of the particular species took several weeks, during which time an invasive case could irreparably damage a victim's face. Wirth and a Harvard associate, Diane McMahon-Pratt, developed a rapid assay using a technology known as nucleic acid hybridization. They found that the DNA of the leishmania parasite would only bond (or "hybridize") with the DNA of its own species and not of another species. Wirth's test involved taking a small biopsy specimen of the skin lesion—skin lesions exist in the early stages of both forms of South American leishmaniasis—and pressing it directly onto a piece of photography paper, which is then dried, blotted with a radioactive form of a known species of leishmaniasis DNA, and developed. When the sample either bonds with the radioactive DNA, as revealed when it gives off light under a special microscope, or fails to bond, the physician knows which form of parasite is causing that patient's infection.

"The materials are extremely stable," said Wirth at the time. "You can put them in an envelope and mail them." This kind of stability is especially important in a test designed for use under primitive field conditions.[63]

Other scientists also have discovered how often ad hoc methods are involved when conducting sophisticated molecular biology, such as nucleic acid hybridization and the newer technique known as PCR (polymerase chain reaction), in the field. One group of scientists from the University of California, San Francisco, found they could isolate DNA for testing simply by boiling their tissue samples, a process that freed DNA. But in Quito, Ecuador, with altitudes of 9,000 feet, their samples boiled dry before getting hot enough to carry out one step of the diagnostic procedure. The scientists adapted by placing a layer of oil over the water to contain it and raise the temperature.[64]

As for environmental control of leishmaniasis, the sand fly vector of leishmania is a much more complicated organism to eradicate than is the schistosome worm. "You can have reasonable sanitation and still have leishmaniasis," says David. "It has much more to do with the trees." In more desert-like parts of the world, insecticides can effectively control the sand flies that carry kala azar. But insecticides are much less effective in South America because of the dense forestation. "If the flies are in the houses, you can spray the house with insecticide," says David, "but if they're outside you can't. You just can't attack the flies effectively as the area of forestation is so extensive and the leaves can shield the sand flies."[65]

The special challenge of trying to eradicate tropical diseases is that public health experts must consider so many factors: the life cycles of vector species, the relationship between vegetation and insecticide, the rainfall, the altitude, the average annual income. Tropical infectious diseases tend to be bound up not only in the pathogen responsible for the disease but also in the life cycle of the vectors and reservoirs, the climate and topography of the endemic area, and the living conditions of the human hosts. More than in almost any other subspecialty of public health, experts in tropical diseases must be truly interdisciplinary.

It is not only in the third world where vector-borne diseases require this multifaceted approach. Some infectious diseases in the United States are every bit as vexing in terms of identifying and interrupting the life cycles of the insect vectors and animal reservoirs involved in their transmission. One of the most dramatic examples is a disease that seemed to arise out of nowhere in the 1970s but that still plagues Americans today: Lyme disease.

Out of the Woods: Lyme Disease

In the summer of 1975 at least five young children in the town of Lyme, Connecticut, were diagnosed with juvenile rheumatoid arthritis, a progressive and debilitating joint disease. The children suffered from swollen knees, aching joints, and difficulty walking. The children's mothers, though, were skeptical of the diagnosis. It made no sense to them that so many youngsters from the same neighborhood were suffering from an illness that was supposedly quite rare.

Two of the women, Anne Mensch and Polly Murray, took their concerns to the state health department. After much persistence, these amateur epidemiologists managed to convince the professionals that something strange was happening in the town of Lyme.[66]

Thanks to the efforts of Mensch and Murray, scientists at Yale University's School of Medicine, led by Allen Steere, began to investigate the apparent outbreak of juvenile rheumatoid arthritis. They pursued every possible explanation for the clustering of cases: a pollutant in the air, radiation from a nearby nuclear power plant, toxins in the water supply. But the most relevant clues were that most victims lived near heavily wooded areas, were affected during the summer months, and seemed to recall having developed a skin rash resembling an insect bite before the symptoms began. To Steere the most likely explanation of the outbreak was an infectious disease spread by an insect vector.

Based on where the bite marks were detected and on a similar disease reported in Scandinavia that was linked to a tick, Steere suspected that a tick was also responsible for transmitting Lyme disease. But it was not until 1977, when one of his patients actually saved the tick he had removed from his body, that the insect was identified. Steere sent the tick to his colleague Andrew Spielman at the Harvard School of Public Health, one of the world's leading authorities on tick-borne illnesses. Spielman said it was a member of a new species that he had identified four years earlier.

Spielman had originally spotted the tick on Nantucket, the moor-filled island off the Massachusetts coast. The first case of what was then called Nantucket fever occurred in 1969 in a summer resident in her sixties. Although at first diagnosed with malaria, an alert investigator at the CDC, George Healy, recognized in the woman's blood sample an obscure protozoan parasite known as babesia. The diagnosis was made and then forgotten.

In the summer of 1973 a friend of the first patient, also a woman in her sixties, came down with symptoms of Nantucket fever that, again, were initially misdiagnosed as malaria. Healy made the diagnosis again and called his friend Spielman at Harvard. As Spielman recalls:

> Knowing that two people were involved, that became interesting. That means this wasn't some strange immunological weakness in the first person; this is now something that's worth looking at.

With the information that the infection was caused by babesia, Spielman could infer that it was transmitted by a tick, since all babesia are tick-borne. He could also infer that the tick most likely lived on rodents. But what tick and what rodent? Spielman and his assistants set off to discover the answer:

Andrew Spielman sweeping for ticks on Nantucket

Our first trip out to Nantucket was by bike, since we didn't have any money. I brought along a technician and his girlfriend, and each of us loaded our bikes with traps, raccoon traps, great masses of things tied on our bikes with strings. On our first trip we caught mice and voles—the first trapping of that kind that I had ever done—and shot jackrabbits. We didn't know what we were looking for. We found immature dog ticks were on voles, and another tick altogether on mice.[67]

The Harvard scientists soon decided they were working with a new species of tick—which they named *Ixodes dammini* after their colleague at Harvard, pathologist Gustave Dammin[68]—that existed on mice only in the larval stage. "So we started looking for the adult stages," says Spielman, "and found them on deer."

Hunting season had begun, and we just picked the ticks off the deer as the hunters brought them in at the checking stations. There was no reason early on why we manned the first checking station; I guess it was just this process of wanting to do an overall survey of everything. Fortunately, these ticks are on a deer during hunting season. We picked off hundreds and hundreds of ticks and reared them out, and compared their larvae and nymphs with the larvae and nymphs we were taking off of mice, [and] found out they were the same thing. We knew at that point that we had a new tick species.[69]

With so much information already in place regarding the transmission of Nantucket fever—more properly known as babesiosis—it was a relatively simple matter to explain the transmission of Lyme disease when Steere sent over his patient's tick in 1977. Then in 1982, five years after the Harvard scientists had identified the insect vector as *I.*

dammini, Willy Burgdorfer, at the Rocky Mountain Laboratories in Montana, finally identified the microbe that caused Lyme disease. It was a tiny spirochete, a corkscrew-shaped bacterium related to the germ that causes syphilis. Luckily, it could be killed with antibiotics such as erythromycin, penicillin, and tetracycline.

Why did Lyme disease appear just when and where it did? The stage had probably been set decades earlier. During the first half of the twentieth century, the densely forested regions of the northeastern United States were cleared to make room for farms and, later, for suburbs. This profound change in the region's ecology cleared the area of white-tailed deer, which had been plentiful. It also cleared the area of the deer's predators. Eventually, some of the forest grew back and the deer returned, but their predators did not. As the deer population grew, so did the population of the deer tick. And the more ticks around, especially in suburban communities where homes are built right up against the woods, the more likely it was that a human being would get bitten by an infected tick.

Once the pattern of transmission was identified, the next step was to try to interrupt it. "As with malaria, control strategies have to be varied to be effective and long lasting," says Spielman.[70] Among the strategies he and his colleagues have tried are the complete removal of all deer from isolated locales (such as Great Island in West Yarmouth, Massachusetts); public education campaigns in endemic regions, so people learn how to avoid ticks and how to recognize Lyme disease; and the judicious use of pesticides.

One kind of pesticide delivery system developed by Spielman and colleagues Thomas Mather and Jose Ribeiro administers a tick-killing chemical through the white-footed mouse, the rodent that harbors the Lyme-carrying tick during its juvenile stage. The scientists stumbled upon their strange-looking trap (in essence, an empty toilet paper roll stuffed with pesticide-soaked cotton balls) almost accidentally. Ribeiro had taken his children to Boston's science museum, where the nocturnal environment exhibit provided cotton balls for the mice to build their nests. He reasoned that even if the cotton were permeated with pesticide, the mice would probably still use it—and the pesticide might kill off the dangerous ticks. In the spring of 1987 the investigators first set out the lethal cotton balls in a field test. Not only did the mice build their nests with it, but the ticks rapidly died off, from

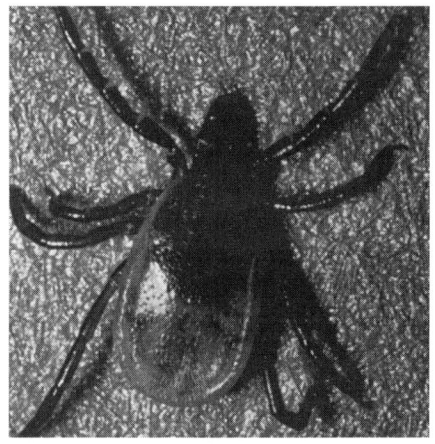

Deer tick, the vector of Lyme disease

an average of about a dozen ticks on each mouse in previous years to less than one per mouse that spring.[71]

Sometimes the least sophisticated, most commonsensical approach to a particular disease—like pesticide-soaked cotton balls set out in empty toilet paper rolls—is the most effective. And once a particular disease is eradicated, life can return to an entire region that had been abandoned because people feared for their health. That is what happened with the tropical infection known as river blindness, which is being controlled thanks to a careful understanding of and attack on the means of transmission of the disease, combined with the discovery of a simple pill and the generosity of an American drug company.

A Cure for River Blindness

The Volta River basin cuts a wide swath across the midsection of the African continent. In centuries past, the river served as a lifeline through the landlocked nations of the savanna. Along its banks, land could be farmed, because the river provided irrigation, transportation, and a continuous supply of rich earth from its muddy bottom.

But by the mid-1970s the land along the Volta had become uninhabitable. Once known for its thriving communities, this region of Africa was becoming known instead for its "macabre processions"—the tragic sight of young children serving as seeing-eye guides for the craggy old men and women who had become blind in midlife. What made these pairings especially brutal to witness was the knowledge that the children, too, would very possibly be struck blind themselves in a matter of a few years.

"This was not a question of a few affected persons," recalls Adetokunbo Lucas, another graduate of the Harvard School of Public Health, who was then head of the tropical diseases research division of the World Health Organization and then worked at the Carnegie Corporation before joining the school's faculty in 1990. The river blindness disease, formally known as onchocerciasis, was wiping out entire communities. As Lucas remembers:

> The intensity of the problem was such that some villages had died, and large areas of arable land along the river banks had been abandoned because flies had literally driven people out. In some of the areas, as many as 15 to 25 percent of the adults were blind. . . . [The disease] was seen not only in terms of health but also as an important barrier to economic development.[72]

Dozens of epidemiologists from all over the world had been studying the mysterious river blindness found in West Africa, the Sudan, and some other parts of Africa since the early 1940s. By the mid-1970s, epidemiologists had mapped out the pattern of disease and they knew how it was transmitted. The blindness was caused by a parasitic worm

45

that infected humans who were bitten by the tiny black fly, which carried the worm larvae. Then the worm went through its life cycle inside the human. Its offspring, called microfilaria, erupted into the skin, causing unbearable itching and, if they infected the eyes, blindness.

Epidemiologists and entomologists had worked out how and where the black flies reproduced—an essential first step if they were to direct an effort to eliminate the disease. It turned out that they were multiplying in waterfalls, river rapids, and other places where the river sped the fastest. But because people don't live near these fast-moving waters, the flies had developed an enormous flight range, up to 500 to 600 kilometers, so they could feed on people to get the warm blood they needed to survive. By contrast, a typical mosquito lives out its entire life span within 1 or 2 kilometers of the spot where it was bred.

Public health strategists adopted a military approach to the problem. In fact, to this day the World Health Organization implements the Onchocerciasis Control Programme on behalf of several international agencies—a successful example of international collaboration in health reminiscent of the smallpox eradication program. They devised local maps, based on reports of fly catches and larva scouting missions, of where the breeding sites were and issued the maps to helicopter bombers with instructions to "go and dump large quantities of insecticide." At first they used a narrow-spectrum insecticide known as Abate, since it caused little damage to untargeted organisms or to people living along the banks of the river. But by the early 1980s the black flies became resistant to Abate.

"At about that time, [the] WHO Tropical Diseases Research Program had also started a program to look for biological agents for vector control," Lucas says. Under this program, researchers had found that a strain of the bacterium *Bacillus thuringiensis* (Bt) was highly potent in killing the black fly larvae but spared many other living things in the river as well as the local fish. An Israeli scientist had earlier discovered this new strain, so it was named *Bacillus thuringiensis var Israeliensis* (Bti). "Bti was never a complete substitute for chemical insecticide because it required very large quantities," says Lucas, "but during that year when this resistance first came in, it served as a stop gap while scientists looked for an alternative insecticide with the same profile as Abate: that is, high efficacy but low risk of damage to the ecosystem."[73]

By 1990, after nearly two decades of operation, the river blindness program seemed to be a success. Ninety percent of the region was declared free of black fly vectors and none of the children under the age of six showed any sign of infection. "People are now migrating back into areas that were abandoned to the flies," says Lucas, "which is evidence of economic recovery and activity in the areas that had been abandoned."

But while the vector control program seemed efficient, it had to be continued for long periods of time, since as many as 18 million people in West Africa still had worms in their bodies—and the worms have a 12-year life span.

In the mid-1980s, scientists found that a drug usually used to treat livestock parasites was also effective against the onchocerciasis microfilaria. The drug, ivermectin, was manufactured by the large American pharmaceutical company Merck. At a public meeting at WHO in Geneva in June 1986, in an unprecedented gesture of goodwill, Merck announced that it would donate ivermectin free of charge to all nations where onchocerciasis is endemic.[74] This was evidence once again of the complexities of solving tropical public health problems—and the enormous changes in the lives of an entire continent that are still possible. But while American know-how again helped vanquish an infectious disease, it met perhaps its most formidable foe at about this same time in the form of a new infection against which all its sophistication was useless: the infection that has come to be known as AIDS.

The Emergence of AIDS and Other New Diseases

In a way we were lucky in identifying AIDS. Because it led to highly unusual symptoms in a clearly defined population, epidemiologists could see quite clearly a pattern in the emerging disease. The earliest AIDS patients were young, previously healthy homosexual men suffering one of two odd conditions: either Kaposi's sarcoma, a rare cancer that until then had only been seen in very old, very sick men, or *Pneumocystis carinii* pneumonia, which until then had hardly ever been seen at all. If the earliest AIDS patients had suffered from more ordinary cancers and infections, it might have taken years to realize that a common underlying infectious disease was causing their different symptoms.

Even in 1981, when the CDC published its first notice of what seemed to be an inexplicable sporadic outbreak of *Pneumocystis* pneumonia in five homosexual men in Los Angeles,[75] epidemiologists suspected something important was happening. Soon, reports started to accumulate of other gay men, in San Francisco and New York, most of them young, who were sick, many with diseases that usually affect the aged.

During those early days of the outbreak, one of the clearest patterns that emerged was that AIDS seemed to be transmitted much the way hepatitis was. In other words, blood and body fluids were probably involved. That likelihood narrowed the search for the microbe that caused the disease to one that could be carried in the bloodstream.

Max Essex, professor of health sciences at the Harvard School of Public Health, was able to narrow it even further. Trained as a veterinarian, Essex had been toiling away in relative obscurity on a cat virus known as FeLV (feline leukemia virus). FeLV is a retrovirus (so named because it reproduces in an opposite, or "retro," way from most other viruses) that causes an AIDS-like immunosuppression in cats. He had

had a difficult time convincing his colleagues of many of his earliest findings.

> I can still recall an NIH workshop on retroviruses in the early 1970s at which I summarized overwhelming evidence that FeLV caused disease in house cats as a horizontally transmitted agent [passed from one cat to another, rather than one that is transmitted vertically, from parent to offspring]. Most of the audience seemed to dismiss the possibility that any retrovirus could naturally cause disease. They assumed that retroviruses were only important as genetically inherited viruses, or "virogenes." The exception was Bob Gallo of the Laboratory of Tumor Cell Biology of the National Cancer Institute [NCI]. He was convinced that retroviruses were not only important as disease-causing agents, but that human retroviruses existed.[76]

By 1975, Essex's lab at the Harvard School of Public Health had generated the first good evidence that a lethal suppression of the immune system—in this case in the cat immune system—could be caused by an infectious agent.[77] In 1978 Robert Gallo's lab at NCI discovered human retroviruses, which Gallo called HTLV viruses (short for human T cell lymphoma virus, one of the cancers linked to HTLV). When the first reports of immunosuppression in gay males trickled in a few years later, both Essex and Gallo were ready. In May 1983 Gallo published two papers in *Science* announcing that he had found HTLV in AIDS patients.[78] In the same issue of *Science*, Essex published two papers with similar results, including one with background information about the related FeLV and the ways in which it suppressed the immune systems of cats.[79]

During the next five years, Essex turned most of his research attention over to AIDS, in particular to outlining the components and characteristics of HIV, the human immunodeficiency virus. With the goal of developing an AIDS vaccine, he set about trying to identify the site on the virus's surface that attaches to the host cell. This site, known as the surface protein, is usually a primary component of any vaccine. The elucidation of the surface protein came relatively quickly; by 1985, Essex and his Harvard colleague, Tun-Hou Lee, had identified two glycoproteins (sugar-plus-protein compounds) on the HIV surface that worked, as Essex puts it, like

Myron (Max) Essex

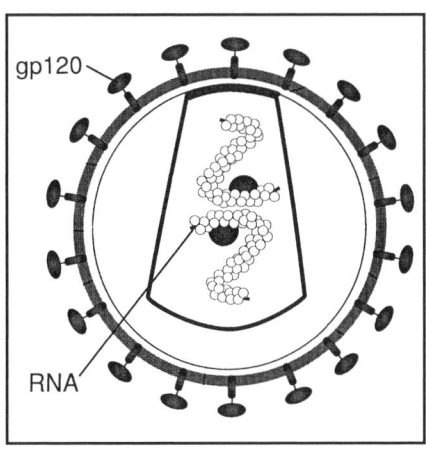

Schematic drawing of HIV virion

"a golf ball on a tee: gp120 is the ball, while gp41 is the tee." One small region of gp120, Essex and Lee found, is the only spot that forms a bond with the host cell that HIV infects.[80]

The relatively small size of this attachment zone helps explain why, more than 10 years after gp120 was discovered, an AIDS vaccine still eludes scientists. Any vaccine tested to date (as well as any HIV infection acquired naturally) leads to the production of what Essex calls "diversionary antibodies." These antibodies, directed at parts of the gp120 molecule that are irrelevant in terms of host cell infection, are unique to HIV and embody the special challenge of developing an AIDS vaccine.

In 1988, to promote an interdisciplinary approach to what was fast becoming the leading public health crisis of the century, Harvard president Derek Bok announced the creation of the university-wide Harvard AIDS Institute, with Max Essex as chairman. The institute was one of the first and most far reaching of its kind, involving experts not only in molecular biology and virology but also behavioral science, biostatistics, clinical medicine, epidemiology, ethics, law, public policy, and veterinary science. "AIDS is bigger, tougher, and more complex than any other disease right now," Essex said at the time. "Other major diseases, however severe, have reached relative states of equilibrium in the world population, while the AIDS epidemic is still growing. We have to attack AIDS before it reaches an equilibrium—otherwise, it would be too devastating."[81]

One of the institute's earliest projects concerned research into a second type of HIV, known as HIV-2, that was discovered a few years earlier by Essex and his colleague Phyllis Kanki. To mount the project, Harvard scientists teamed up with researchers in France (at the University of Tours and the University of Limoges) and Senegal (at the University of Dakar). They travel to Senegal four or five times a year, working out of the handful of flat-roofed stucco buildings that constitute the laboratory of the Hôpital le Dantec in Dakar. A typical scene includes families bringing food to their sick while, outside, others family members sit or do laundry, children play, and animals graze nearby. Inside, in several small rooms, health care workers examine patients and test them for infections ranging from strep throat to syphilis to AIDS. In another room, workers are performing tests on patients' T-cells.

Confronting the Biological Environment

49

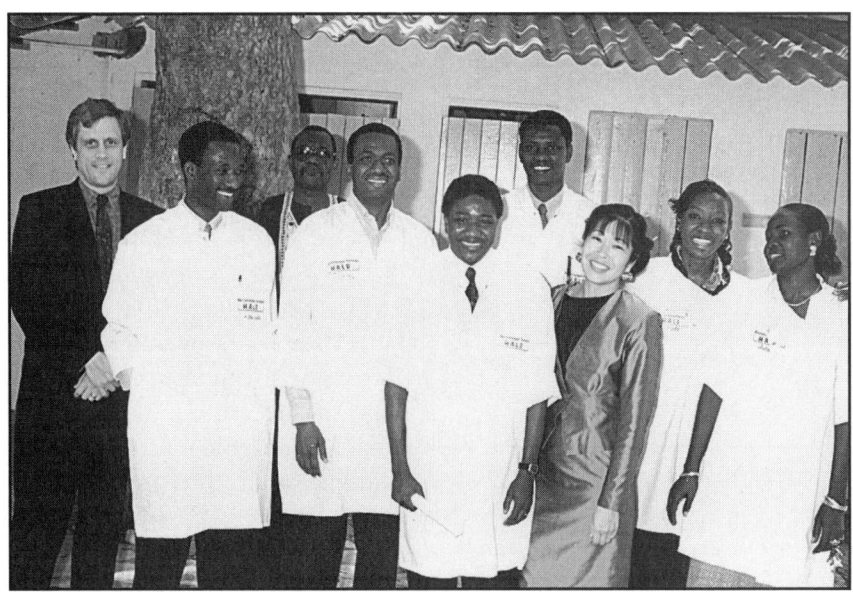

*Phyllis Kanki (third from right) and Richard Marlink (far left)
with colleagues in Senegal*

The Boston contingent packs dozens of boxes for each of these forays, including up to 10,000 AIDS kits at a time. They take along "syringes, counter-top centrifuges, plastic gloves, drugs, and all the mundane, obvious things," says Essex, as well as ". . . condoms donated by manufacturers. And, of course, someone always carries a computer back and forth in a knapsack."[82]

Most of the research in Senegal focuses on HIV-2, the most common form of HIV infection in West Africa. (HIV-2 infection occurs only rarely in the United States and Europe.) Unlike HIV-1, which is responsible for the worldwide pandemic of AIDS, HIV-2 causes a far more languorous infection. People infected with HIV-2 show fewer immune system abnormalities, develop fewer opportunistic infections, and are more likely to live longer before AIDS develops.

A clear demonstration of HIV-2's slow clinical course came in 1994, when Kanki and Richard Marlink of the Harvard AIDS Institute completed an eight-year study of Senegalese prostitutes. Among the 574 female sex workers they studied in Dakar, Kanki and Marlink found that, even five years after infection, none of those infected with HIV-2 had developed symptoms of AIDS. In contrast, among the prostitutes infected with HIV-1, one-third were sick with AIDS within five years.[83]

The following year Kanki and her associates reported on continued observations of this group of prostitutes, concluding that infection with HIV-2 reduced a female's chance of acquiring HIV-1 infection by approximately 70 percent. Significantly, the prostitutes in Kanki's study were still contracting gonorrhea at a high rate, so their low rates of HIV-

1 infection were not due to changes in behavior. "Clearly, HIV-2 is triggering a resistance response that in some way protects against HIV-1 infection," Kanki explains. "These findings suggest a new avenue for research into fine-tuning that natural protection, and delivering it in the form of a safe vaccine." She emphasizes that no one is talking about using HIV-2 itself as a live vaccine—only about using it as a model for developing a safe vaccine with the appropriate protective components.[84]

Investigators at the Harvard AIDS Institute have also focused on another subset of HIV: forms of HIV-1 known as non-B subtypes. According to Essex, the B subtype of HIV-1—the most common form found in the United States—is not easily transmitted through heterosexual intercourse. But the non-B subtypes (of which there are at least five that are more common in Africa and Asia) are especially suited to heterosexual transmission. Indeed, according to Essex, the risk of HIV transmission through vaginal sex is 40 times higher for a non-B subtype of HIV than for the B subtype.[85]

The countries where the AIDS epidemic is growing most rapidly today, India and Thailand, are places where new non-B strains of HIV are predominant. "The E subtype from Africa went into Thailand, and the C subtype went into India," Essex says, which "caused massive heterosexual epidemics. So the question is, when will we in the United States get a new epidemic based on one of these other forms, and when we do, will it infect many more heterosexuals? Almost certainly."[86]

HIV is the most notorious of a group of pathogens that have come to be known as emerging microbes. A long line of supposedly new infectious diseases has been fodder for news accounts for decades, beginning with Lyme disease in 1975 and then Legionnaire's disease, toxic shock syndrome, Hantavirus, Ebola virus, new strains of cholera, drug-resistant pneumococcus, and group A streptococcus (which the tabloids dubbed "flesh-eating bacteria"). Even infections that the experts had thought were long since vanquished—most prominently tuberculosis, the nineteenth-century scourge that had been quelled by antibiotics in the 1950s—are back in the headlines as resurgent health threats.

All this microbial activity caught many experts by surprise. Not so very long ago, Surgeon General William Stewart boldly stated that the war against infectious diseases had been won. When he made that declaration, in 1969, all signs did indeed point to victory. Vaccines had virtually eliminated smallpox and polio and antibiotics had defanged others, such as syphilis and tuberculosis.

Stewart was not alone in precipitately declaring victory over the microbial world. Andrew Spielman remembers an evening in the 1950s when his graduate adviser at Johns Hopkins took him to a Baltimore bar for a beer and a talk. He told Spielman he was feeling guilty about encouraging the young man to enter the field of public health entomology, which he predicted would be eliminated in a matter of years thanks to the advent of DDT. Forty years later Spielman looks back on a busy career at the Harvard

School of Public Health, usually juggling investigations into three or four vector-borne infectious diseases at a time.

How could the experts of the 1950s and 1960s have been so wrong, so wildly optimistic in their predictions? Richard Levins of the Harvard School of Public Health has asked himself this question. They were wrong, he says,

> because people were working from too narrow a theoretical basis. People simply extrapolated from tendencies that they had seen over the previous 100 years: smallpox was down, TB was down, malaria was down. But if you look more broadly, across other animal and plant diseases, you will see that diseases appear and disappear. And if you look at the whole sweep of human history, you'll also see that in periods of major social and environmental change, we also see changes in health.[87]

Of course, inaccurate predictions can create problems that arise out of sheer complacency when people mistakenly let down their collective guard. That is what, no doubt, happened with malaria. The attitude toward malaria in the 1960s—that it was a soon-to-be-extinct disease—led to a terrible vacuum in treatment options when the eradication techniques backfired. A similar attitude toward bacterial infections in the 1980s led to another treatment vacuum as more and more bacteria became resistant to one or more antibiotics. As the drugs on the market became impotent, there were no good alternatives in the research pipeline because so many scientists had assumed that antibiotic development was an outmoded career.

The CDC recently estimated that between 1982 and 1991 the rate of drug-resistant tuberculosis more than doubled in the United States. The situation was worst in impoverished urban areas, but drug-resistant strains of tuberculosis, in which death rates were as high as 75 percent, were reported in at least 17 states. In addition, by the early 1990s about one in every 10 strains of pneumococcus (*Streptococcus pneumoniae*) was resistant to every known antibiotic except one, vancomycin. If those strains became resistant to vancomycin, too, physicians would have no ammunition left to combat many infections with pneumococcus, the most common cause of bacterial pneumonia and the cause of about one-half of all childhood ear infections, as well as many cases of meningitis, sinusitis, and bronchitis.

To keep the international public health community from letting down its guard in relation to infectious diseases, the Harvard School of Public Health formed a Working Group on New and Resurgent Diseases in 1991. The group was charged with considering the issue from a truly multidisciplinary perspective. Included on its membership roster were experts in the fields of biomathematics, clinical infectious diseases, ecology, epidemiology, philosophy, sociology, and systems ecology.[88] They concluded that most of the new diseases that have emerged in the past few decades can be traced to some change in the environment, either natural or caused by humans. "There is no 'Andromeda strain' dropping from the sky," says Spielman, a member of the working group. "Any pathogen we encounter will either be one already existing or one with a slight modification."[89]

When the Hantavirus emerged in the American Southwest in 1993, for instance, epidemiologists eventually traced its appearance to a virus that had been harbored for years by a rodent known as the deer mouse. After an unusually wet spring, which provided ample food for the deer mouse, the rodent population grew significantly and moved closer to people's homes, increasing the likelihood that humans would be exposed to the virus the mouse carried.

Although new diseases do occasionally emerge as a result of climate change or other natural changes in the environment, as did the Hantavirus, more often they can be traced to some change in human behavior.[90] The human action changes the environment so that an existing microbe can take up residence in a new niche, in proximity to a new population. Once this happens, a disease that has been around for a long time—during which time it has remained relatively harmless—captures our attention because it starts appearing in a new population, or a new geographic region, or a new host species. Among the most important of these human activities are the following:

- *Globe trotting by infected individuals.* With international airplane travel, people can cross continents in a matter of hours; if they are infected, they can bring microbes to unfamiliar sites long before their first symptoms appear.
- *Building homes near wildlife habitats.* As housing construction moves into wooded or forested areas, destroying animals' natural habitats, humans are brought into close contact with animals who approach their neighborhoods in search of replacement sources of food and shelter.
- *Urbanizing sparsely populated areas.* This brings together people from different regions, often in squalid living conditions that foster the transmission of contagious diseases.
- *Changing the marine ecology.* Waters contaminated with nitrogen-rich wastewaters, fertilizers, and acid rain encourage the proliferation of coastal seaweed and algae, which can harbor the bacteria of cholera and other water-borne diseases.
- *Changing agricultural economies.* The introduction of new crops brings new crop pests, and the microbes they harbor, close to farming communities, subjecting agricultural workers to new diseases.
- *Transporting microbes from one region to another.* Microbes can be transported, for instance, inside mosquitoes or "wharf rats" that hitch a ride on ships moving cargo between hemispheres.
- *Building roads into rain forests.* This brings humans into contact with insects and animals that harbor unfamiliar microbes.

New diseases have emerged in these ways for centuries. In the sixteenth century, for example, European explorers brought smallpox to the Americas. More than 100 years later, the slave trade brought yellow fever from Africa to the New World. Sailing ships were the ideal conveyance for the *Aedes aegypti* mosquito, which carries the yellow fever virus, because drinking water was stored in barrels. *Aedes aegypti* will breed only in artificial containers; it will not lay its eggs in water with a muddy or sandy bottom, such as lakes or ponds. When ships began transporting African slaves to the Caribbean, the open water barrels on board were perfect breeding grounds for *Aedes aegypti* larvae. Once the boats docked in the Caribbean, the warm climate allowed the tropical mosquitoes to thrive. At that point, all it took was for some Europeans with yellow fever to come into the vicinity, and the mosquitoes in the Caribbean were infected—and able to pass on the yellow fever virus to unexposed nonimmune natives. In this way, yellow fever made its entry to the New World in 1648 and has continued to plague this hemisphere with sporadic outbreaks ever since.

A more recent twist to this story occurred in 1984, when a different, more aggressive species of mosquito, the so-called Asian tiger mosquito, was inadvertently imported from Japan into Texas in a shipment of used tires. The few ounces of rain water that accumulated inside the tire rims was enough for the mosquito to lay and hatch its larvae, releasing thousands of tiger mosquitoes into the Texas port. Since then, the tiger mosquito has spread to 18 states, mostly in the southwestern part of the country, and has actually replaced the *Aedes aegypti* population in many regions. Significantly, the tiger mosquito can carry not only the virus that causes yellow fever but also the virus that causes two related diseases: dengue fever, which is relatively mild, and the more deadly dengue hemorrhagic fever. Many experts consider dengue fever to be the most likely candidate for the next American plague. No native cases have yet arisen in the United States, but it seems only a matter of time. Now that the mosquito vector has arrived, only a few sick people are needed for dengue to take hold.

Some new diseases have gained a foothold because of changes in modern agriculture. The deadly Argentinean hemorrhagic fever, for instance, emerged after World War II, when the pampas grasses of Argentina were cleared to make way for new farms in the northwestern provinces of the country. When the tall grasses disappeared, so did predators of the *Calomys*, a rodent that carries a pathogen known as the Junin virus. During late summer (which is February in Argentina) to early winter, the *Calomys* population increased. The human population increased, too, during those months, as migrant farmers moved in for the harvest. The result was that each year there was an outbreak of a terrible infection that killed up to 20 percent of its victims.

The Harvard working group has taken special note of changes in marine ecosystems that have led to the emergence of several new strains of cholera around the world. "Sewage and fertilizer pouring into marine ecosystems, overharvesting of fish and shellfish and the loss of wetlands, combined with climatic changes, have conspired to cause a worldwide explosion of coastal algal blooms around coastal regions, providing a rich environment for diverse communities of microorganisms," the group writes.[91] These microorganisms become a health threat to humans who eat the fish, mollusks, and crustaceans infected with them.

In many underdeveloped countries throughout the world, war and famine provide the ideal conditions for the reemergence of louse-borne diseases such as epidemic typhus. As an editorial in *The New England Journal of Medicine* pointed out recently:

> Tens of millions were infected with typhus during and after World War I, and millions died. Typhus continues to afflict humans in epidemics that the developed world largely ignores. We should consider carefully how to prevent this disaster from occurring in Bosnia, the republics of the former Soviet Union, African nations with major internal conflicts, and other places where living conditions, hygiene, or sanitation have deteriorated and thus favor person-to-person spread of disease by body lice.[92]

The study of emerging diseases makes clear the interrelationship between the biological and the physical environments. This interrelationship brings the study of microbes full circle: when bacteriology was in its infancy, some public health officials believed that physical causes of disease were a thing of the past. Today, what emerging diseases demonstrate is that the natural world must be addressed as a whole. Things that occur in the microbial world affect the environment; things that occur in the environment affect microbes; and each system in turn affects the public's health.

The ideal approach to protecting the public's health involves a wide range of approaches: biological, physical, and social. "Other organisms develop very nonspecific resistances that confer protection against a range of threats," the Harvard working group writes. "For human beings, measures could be taken to boost the body's defenses in general. These would include good nutrition, pollution control, biodiversity for vector control and social arrangements that ensure these measures reach the entire population."[93] In other words, the biological perspective of disease must embrace the wisdom, and the tactics, implicit in the physical perspective—a perspective to which we now turn.

Notes

1. Paul de Kruif, *Microbe Hunters*. New York: Harcourt Brace, 1926, p. 122.
2. Ibid., p. 150.
3. D. H. Leonard, "The bacillus of tuberculosis and the aetiology of tuberculosis—is consumption contagious?" *Journal of the American Medical Association*, 1884, vol. 2, p. 463.

4. William H. McNeill, *Plagues and Peoples.* Garden City, N.Y.: Anchor Press/Doubleday, 1976, pp. 273–274.

5. Laurie Garrett, *The Coming Plague: Newly Emerging Diseases in a World Out of Balance.* New York: Farrar, Straus and Giroux, 1994, p. 403.

6. Elizabeth Fee and Dorothy Porter, "Public health, preventive medicine, and professionalization: Britain and the United States in the nineteenth century." In E. Fee and R. H. Acheson, eds., *A History of Education in Public Health: Health That Mocks the Doctors' Rules.* Oxford and New York: Oxford University Press, 1991, p. 33.

7. Robert S. Desowitz, *The Malaria Capers: More Tales of Parasites and People, Research and Reality.* New York: W. W. Norton, 1991, p. 151.

8. "Malaria" in J. R. M. Kunz and A. J. Finkel, eds., *The American Medical Association Family Medical Guide.* New York: Random House, 1987, p. 579.

9. McNeill, *Plagues and Peoples*, p. 101.

10. Fitzhugh Mullan, *Plagues and Politics: The Story of the United States Public Health Service.* New York: Basic Books, 1989, p. 125.

11. Andrew Spielman, personal interview, May 1994.

12. Ibid.

13. Desowitz, *The Malaria Capers*, p. 62.

14. Gordon Harrison, *Mosquitoes, Malaria and Man: A History of the Hostilities Since 1880.* New York: E. P. Dutton, 1978, p. 219.

15. Spielman, personal interview. A small amount of DDT was typically administered in proportionately much larger amounts of talcum. Consequently, use of DDT often produced dust-like clouds.

16. McNeill, *Plagues and Peoples*, p. 282.

17. Mullan, *Plagues and Politics*, p. 125.

18. Spielman, personal interview.

19. Garrett, *The Coming Plague*, p. 48.

20. John M. Karefa-Smart, "The challenge of unfinished tasks," *Harvard Public Health Alumni Bulletin*, Jan. 1964, vol. 21, no. 1, p. 3. The article is from the opening speech given by Karefa-Smart at the Fifth Conference of the Industrial Council for Tropical Health, Harvard School of Public Health.

21. Data on the prevalence of malaria worldwide come from the National Institute of Allergy and Infectious Diseases, disseminated at the third annual meeting of the institute's International Centers for Tropical Disease Research, April 27–29, 1994.

22. Andrew Spielman, Uriel Kitron, and Richard J. Pollack, "Time limitation and the role of research in the worldwide attempt to eradicate malaria," *Journal of Medical Entomology*, 1993, vol. 30, no. 1, p. 6.

23. Harry F. Dowling, *Fighting Infections: Conquests of the Twentieth Century.* Cambridge, Mass.: Harvard University Press, 1977, p. 209.

24. Peter Radetsky, *The Invisible Invaders: The Story of the Emerging Age of Viruses.* Boston: Little, Brown, 1991, p. 108.

25. Catherine Drinker Bowen, *Family Portrait.* Boston: Little, Brown, 1970, pp. 241–242.

26. Ibid., p. 235.

27. The account of Philip Drinker's invention of the iron lung comes primarily from two sources: his sister's book, Bowen, *Family Portrait*, pp. 237–245, and Jean A. Curran's *Founders of the Harvard School of Public Health* (New York: Josiah Macy, Jr., Foundation, 1970, pp. 163–165).

28. Bowen, *Family Portrait*, p. 240.

29. Ibid., p. 243.

30. Ibid., p. 245.

31. Radetsky, *Invisible Invaders*, p. 108.

32. Thomas Weller, personal interview, May 1994.

33. Ibid.

34. Ibid.

35. Radetsky, *Invisible Invaders*, p. 116.

36. Eli Chernin, *Tropical Medicine at Harvard: The Weller Years, 1954–1981.* Boston: Harvard School of Public Health, 1985.

37. The most active contributors to polio vaccine research were David Bodian, Howard A. Howe, and Isabel Morgan Mountain at Johns Hopkins; Dorothy Horstmann at Yale; Jonas Salk at the University of Pittsburgh; and Albert Sabin at the University of Cincinnati. The story of their collective work has been told many places, but perhaps most thoroughly by Greer Williams in *Virus Hunters* (New York: Alfred A. Knopf, 1959).

38. The battle for the Salk vaccine trial, including the struggle between Basil O'Connor of the National Foundation for Infantile Paralysis (who wanted to speed things up) and Thomas Francis of the University of Michigan (Salk's mentor, who wanted to slow things down), is vividly captured by Williams, *Virus Hunters*, pp. 295–320.

39. Williams, *Virus Hunters*, pp. 304-305.

40. Dowling, *Fighting Infections*, p. 215.

41. Williams, *Virus Hunters*, pp. 313-320.

42. Dowling, *Fighting Infections*, p. 215.

43. Ibid., p. 218.

44. Williams, *Virus Hunters*, p. 348.

45. Donald Hopkins, "Smallpox," *Harvard Public Health Alumni Bulletin*, Winter 1976, vol. 33, no. 1, p. 7 (transcript of a lecture).

46. Ibid., p. 7.

47. The line of succession of CDC directors who were also alumni of the Harvard School of Public Health spanned the years 1961 through 1989: James Lee Goddard (who got his M.P.H. from Harvard in 1955 and was CDC director from 1961 to 1966); David J. Sencer (M.P.H., 1958; CDC director from 1966 to 1977); William H. Foege (M.P.H., 1965; CDC director from 1977 to 1983); James O. Mason (M.P.H., 1963, D.P.H., 1967; CDC director from 1983 to 1985); and Donald R. Hopkins (M.P.H., 1970; CDC acting director from 1985 to 1989).

48. Hopkins, "Smallpox," p. 8.

49. Ibid., p. 9.

50. Ibid., p. 9.

51. Ibid., pp. 7–8.

52. Robin Marantz Henig, *A Dancing Matrix: Voyages Along the Viral Frontier.* New York: Alfred A. Knopf, 1993, p. 35.

53. John Maurice, "Virus wins stay of execution," *Science*, 1995, vol. 267, p. 450.

54. Ibid., p. 450.

55. For historical background see Charles Siebert, "Smallpox is dead: long live smallpox," *The New York Times Magazine*, Aug. 21, 1994, p. 33. For a news report of the October 1994 vote, see David Brown, "Old enemy of mankind faces execution in June: global assembly must affirm smallpox's fate," *The Washington Post*, Oct. 30, 1994, p. A3.

56. Maurice, "Virus wins stay of execution," p. 450.

57. Richard Anthony, "Global balance," *Harvard Public Health Review*, Winter 1991, vol. 2, no. 2, p. 27.

58. John David, personal interview, May 1994.

59. Greg Folkers, "New vaccine strategy promising against schistosomiasis," *NIAID News*, April 27, 1994, p. 1.

60. Data on schistosomiasis come from the National Institute of Allergy and Infectious Diseases, disseminated at the third annual meeting of the institute's International Centers for Tropical Disease Research, April 27–29, 1994.

61. Donald Harn, personal communication, March 21, 1996.

62. Desowitz, *The Malaria Capers*, p. 34.

63. "Diagnosing tropical disease: Dr. Dyann Wirth's important new test," *Health Sciences Report: News from the Harvard School of Public Health*, May/June 1983, p. 9.

64. Jeffrey Norris, "UCSF researcher brings molecular biology to Latin America," *UCSF News*, Sept. 15, 1994, p. 3.

65. David, personal interview.

66. Joseph Wallace, "Solving the mystery of Lyme disease." In *The World Book Health & Medical Annual, 1989.* Chicago: World Book, Inc., 1988, pp. 40–53.

67. Spielman, personal interview.

68. Ironically, Gustave Dammin, an emeritus professor of pathology at the Harvard Medical School and a lecturer in tropical public health at the Harvard School of Public Health, was bitten in 1981 by the very tick that bears his name and contracted Lyme disease.

69. A controversy erupted in the late 1970s, and remains active to this day, as to whether Spielman and his colleagues identified a new distinct species or merely a northern version of the southern deer tick, *Ixodes scapularis*. Scientists at Georgia Southern University have published findings that the two species are virtually indistinguishable, but the Harvard scientists contest those findings. The most recent account of this argument was summarized by Stephen M. Rich, Diane A. Caporale, Sam R. Telford III, et al., in "Distribution of the *Ixodes ricinus*-like ticks of eastern North America," *Proceedings of the National Academy of Sciences*, 1995, vol. 92, pp. 6284–6288.

70. Harvard School of Public Health, "Once bitten," *Health Sciences Report*, Nov. 1988, p. 1.

71. Ibid., p. 2.

72. Adetokunbo Lucas, personal interview, May 1994.

73. Ibid.

74. Erik P. Eckholm, "Conquering an ancient scourge," *The New York Times Magazine*, Jan. 8, 1989, pp. 20–27.

75. M. S. Gottlieb, H. M. Schanker, P. T. Fan, et al., "*Pneumocystis* pneumonia—Los Angeles," *Morbidity and Mortality Weekly Reports*, 1981, vol. 30, pp. 250–252.

76. Max Essex, "The HIV-1 vaccine dilemma: lessons from the cat," *The Journal of NIH Research*, March 1995, vol. 7, p. 37.

77. Max Essex, "Horizontally and vertically transmitted oncornavirus of cats," *Advances in Cancer Research*, 1975, vol. 21, p. 175.

78. For their early work on AIDS, both Gallo and Essex, along with Luc Montaigner, were awarded the Albert Lasker Medical Research Award, the U.S.'s highest medical award. Thus, they joined Alice Hamilton and many others for groundbreaking research.

79. Radetsky, *Invisible Invaders*, p. 328.

80. Max Essex, "Confronting the AIDS vaccine challenge," *Technology Review*, Oct. 1994, p. 25.

81. Paula Brewer, "Harvard fights back: leading scientists band against AIDS," *Harvard School of Public Health Alumni Bulletin*, Oct. 1988, p. 14.

82. Harvard School of Public Health, "Tracking AIDS in Africa," *Health Sciences Report*, May 1988, p. 2.

83. Harvard AIDS Institute, "People infected with HIV-2 found to have longer-term survival than those infected with HIV-1," press release, Sept. 9, 1994.

84. Paula Brewer, "People infected with second AIDS virus found to have natural protection against HIV-1," press release, Harvard AIDS Institute, June 16, 1995.

85. Craig Horowitz, "Has AIDS won?" *New York*, Feb. 20, 1995, vol. 28, no. 8, p. 37. See the published study, Luis E. Soto-Ramirez et al., "HIV-1 Langerhans' cell tropism associated with heterosexual transmission of HIV," *Science*, 1996, vol. 271, pp. 1291–1293.

86. Ibid.

87. Ellen Hoffman, "Controlling infectious diseases," *Harvard Public Health Review*, Spring 1994, p. 9.

88. Working group members are Richard Levins, Tamara Awerbuch, Irina Eckardt, Paul Epstein, Najwa Makhoul, Cristina A. de Possas, Charles Puccia, Andrew Spielman,

and Mary E. Wilson. Founding member Uwe Brinkmann, an international health epidemiologist, died suddenly in 1993 while conducting field work in Brazil.

89. Hoffman, "Controlling Infectious Diseases," p. 12.

90. A fuller explanation of the emergence of new diseases can be found in Henig, *A Dancing Matrix*.

91. Richard Levins, Tamara Awerbuch, and Uwe Brinkmann, et al., "The emergence of new diseases," *American Scientist*, Jan./Feb. 1994, vol. 82, pp. 59–60.

92. David H. Walker and J. Stephen Dumler, "Emerging and reemerging Rickettsial diseases," *The New England Journal of Medicine*, 1994, vol. 331, p. 1652.

93. Levins, Awerbuch, and Brinkmann, et al., "The emergence of new diseases," p. 60.

*Confronting
the Biological
Environment*

 **3**

Confronting the
Physical Environment

Cleaning up the environment is, seemingly, the most straight-forward task of public health. It has a beginning, middle, and end: you detect a health problem, figure out what in the environment is causing it, get rid of it, and then see an improvement in health. Yet in public health, nothing is truly straightforward. This chapter relates a series of tales about confrontations with the physical environment that reveal how the push and pull of competing interests makes the paths of each of these adventures winding and complicated.

- Workplace safety, which pits the industries that lifted the nation out of the Depression against their own workers, and ultimately delivers the same unhappy news to both employer and employee: that industrial growth often comes at a significant cost to a community's health.
- Radiation health, which also pits industry against exposed workers and against environmentalists and which has caused a rift in the academic community about how much exposure is too much.
- Air pollution, whose dark side was first uncovered in the 1950s after deadly "smogs" descended on two industrial cities and which, despite improvements over the past 25 years, persists as a major preventable source of disease and premature death.
- Water pollution, especially as seen in the town of Woburn, Massachusetts, where a group of biostatisticians concluded that contaminated well water was causing an epidemic of childhood leukemias—and found themselves embroiled in a battle not only against the industrial source of the contamination but also other public health professionals.
- Fluoridation of the water supply, endorsed by every medical and public health organization in the nation as a way to prevent dental cavities but undermined by powerful community organizations that saw it as a threat to their health and independence.

- The Delaney clause, the U.S. Congress's effort to assure the purity of foods, drugs, and other products containing chemical additives, which outlawed any measurable known carcinogen and which has become increasingly unrealistic as the definition of "measurable" became smaller and smaller, down to one part per billion.
- Medical intervention with the drug DES, which was considered harmless in the 1950s and 1960s but was proven otherwise 20 years later, and the battle over DES and other estrogen-like compounds in the meat supply.
- Lead exposure, a pervasive source of obvious and not-so-obvious chronic problems in childhood and the many forces fighting to establish a consensus on what level should be considered "safe."

In each of these stories, public health officials have a common goal: to balance the risks of exposure with the costs of cleanup. Generally, they have had to reach some compromise about how many airborne particulates, how much lead, and how many pesticide residues are to be tolerated. Given the infeasibility of a totally pristine physical environment, the real art of public health in the late twentieth century is to ensure that the physical environment is at least safe enough.

Today's environmental health professionals are far more sophisticated, and far more scientifically grounded, than the sanitarians of a century ago, but they often find themselves promoting the same basic goals. Col. George E. Waring, Jr., for instance, a leader in the sanitationist movement of the nineteenth century, mounted a campaign that involved the introduction of some very modern hygienic techniques: state-of-the-art drains, water closets, sinks, sewers, and cesspools. Yet he promoted these methods from an outdated miasmatic perspective, as a way to fight what he called "zymotic diseases" or "filth diseases" that arose from exposure to "sewer gases." Through all his pronouncements, Col. Waring was never concerned about direct exposure to human waste, which is now known to be the main route of transmission for many leading pathogens of his time. He considered human feces an aesthetic problem only, no more dangerous to health than "decaying potatoes, bad smells, untidy streets, or trash in vacant lots."[1]

Still, whether cleaned up for aesthetic or hygienic reasons, human wastes, rotting potatoes, and trash in vacant lots were far easier to sweep away than are some of the pervasive environmental toxins discussed in this chapter, toxins that public health scientists have been grappling with for decades.

Workplace Risks

In the first half of the twentieth century, jobs could kill. The working conditions in a wide range of industries were abysmal; just showing up for work every morning could be a dangerous act. As the United

Alice Hamilton

States became increasingly industrialized, the list of common occupational diseases grew. By the 1940s, the list included mercury poisoning among felt-hat workers (which caused a neurological disease that led to the old expression "mad as a hatter"), pneumoconiosis among coal miners, and pulmonary tuberculosis among granite cutters.

One of the first people to develop a scientific method for demonstrating the connection between the work one did and the ills one suffered was a pioneer named Alice Hamilton. In an era when the handful of college-educated women in America went into "ladies' work," like child welfare or teaching, Alice Hamilton jumped feet first into the traditionally male profession of medicine—and at the traditionally male enclave of Harvard University. Hamilton was the first woman on the faculty of the Harvard Medical School, which she joined in 1919, and the first on the School of Public Health's faculty, where she held a joint appointment until her retirement in 1935.[2] Indeed, she was the first woman appointed to a faculty of Harvard University. She plunged into her work with relish, according to biographer Barbara Sicherman:

> In her search for poisonous dusts, [Hamilton] jumped on table tops, interviewed workers in saloons, climbed dangerous catwalks, and descended deep into mine shafts, clad in workers' overalls. She also slept in unlocked mining shacks, was mistaken for a prostitute, and at least once wangled her way in through the back door of a large manufacturing company that feared her presence.[3]

Hamilton helped initiate a major study of lead poisoning, which demonstrated that lead was more toxic when inhaled than when swallowed—a finding that ran counter to the conventional wisdom of the time.[4] She collaborated on several other occupational health studies, including those on mercury poisoning in the felt-hat industry and in quicksilver mines.[5]

Yet despite the prominence she attained in the field of industrial hygiene and the vision she brought to her work at Harvard, Hamilton always felt a tenuous, uncertain connection to the university. In part, this was because she insisted on only a half-time position so that she could maintain her research activities in the field. But for whatever reason, Harvard never made Hamilton feel especially welcome. The often-repeated story is that she was invited to join the faculty with three caveats: first, she could not become a member of the all-male Harvard Faculty Club (she could not even set foot inside its doors); second, she could not sit on the platform with the rest of the faculty during commencement ceremonies; and third—and perhaps most stinging of all— she was not entitled to season tickets to Harvard football games like every other faculty member.[6]

As a pioneering woman in a man's world, Hamilton was a victim of an enormous amount of sexism. Despite her nationally recognized contributions to the field of occupational health and her acknowledged role as the "mother of industrial hygiene," she was still only an assistant professor when she retired from Harvard after 16 years.

In 1947, 12 years after her retirement, Hamilton won the Lasker Award, a sign of national recognition for outstanding medical research that is often regarded as the American counterpart of the Nobel Prize in medicine. The Lasker Award jury cited her accomplishments in identifying a long list of occupational ills: "phossy jaw" in the match industry,[7] chemically induced insanity among workers at rayon plants in the textile industry, TNT poisoning during World War I, and illnesses caused by organic solvents such as benzol and carbon disulfide.[8]

Around the time that Hamilton was making her mark on the fledgling field of industrial hygiene, her colleagues were coming to an important realization about the common approaches they shared with experts in other public health disciplines. As Theodore F. Hatch, an alumnus of the Harvard School of Public Health, wrote in 1948, when he was research director of the Industrial Hygiene Foundation:

> The same basic approach which had been employed so successfully in the study and control of infectious disease could be applied similarly to the control of industrial diseases. That is to say, the causative agent was found in the environment; its effect upon man was studied by epidemiological methods and the primary measures of prevention were based upon control of the environment rather than the control of man.[9]

As the 1900s progressed, the nature of occupational hazards evolved. Whereas earlier in this century the biggest dangers came from exposure to chemicals and toxic dust, by the 1950s the risks were changing. Industrial hygienists have gradually become conversant with health problems associated with on-the-job exposure to radiation, high-decibel equipment, dangerous infectious organisms, and more subtle hazards such as repetitive stress injury and eye strain from working at computer monitors.

In fact, many of the public health risks that are most familiar to us today, such as exposure to radioactive radon gas in homes, first came to light because of occupational exposures at levels so high that the health effects were obvious. Like canaries in a coal mine, workers often act as sentinels for the rest of us when exposure is on its way to becoming dangerously high. No case illustrates this reality more clearly than that of the radioactive element known as radium.

Radium Dial Painters and Other Hazards of Radioactivity

When radium was discovered by Marie and Pierre Curie in 1898, it seemed to be nearly magical. It took the Curies years to isolate it from pitchblende ore, working in a leaking wooden shed, laboriously pulling out tiny specks of the radioactive metal. After four years and about a ton of pitchblende, they had their precious reward: a single gram of radium.

Later, when people began working with x-rays and spending their days up to their wrists in radioactive fields, fears arose about the health effects of radiation. Radium workers and early radiologists started to develop sores and ulcers on their hands; many of the sores became cancerous. Walter Dodd, the first chief of radiology at the Massachusetts General Hospital, died in 1916 of complications of skin cancer arising in a radiation-induced ulcer. Madame Curie herself, as well as her daughter Irene, died of anemia and leukemia, which experts today attribute to their exposure to radiation.

In the United States one of the earliest clues about the hazards of radium was the health of employees in factories producing glow-in-the-dark watches. The numbers on the watch dials were painted with radium paint. Beginning in the 1920s, the dial painters from one New Jersey plant, who were mostly women, started showing up in their dentists' offices with painful and disfiguring decay of their jawbones, gums, and teeth.

A group of scientists from the Harvard School of Public Health first pieced together an explanation for these strange dental com-

Confronting the Physical Environment

Turn-of-the-century radium workers

Summary of Blood Findings in Twenty-two Radium Workers as Contrasted with the Blood Findings in Normal Persons

Blood Findings	Radium Workers %	Normal Persons %
Red blood cells		
below 4 million	6	0
between 4–6 million	75	100
above 6 million	19	0
White blood cells		
below 7 thousand	27	0
Blood films		
Polymorphonuclear cells		
between 60–72%	55	100
below 60%	45	0
above 35%	27	0
Lymphocytes		
between 20–25%	18	100
above 25%	64	0
above 35%	27	0
Mononuclear cells		
between 3–8%	41	100
above 8%	59	0
Abnormal red cell forms	36	0
No blood abnormalities	0	100

From the Drinkers' report on radium, "Necrosis of the Jaw in Workers," *Journal of Industrial Hygiene,* 1925.

plaints. Cecil Drinker, a pioneer in industrial medicine and at the time head of the school's Department of Industrial Hygiene, worked with colleagues Katherine Drinker (his wife) and William Castle to determine what was making the workers sick. Was it something in the air in the factories? A poisonous fume, perhaps, or an infectious agent that they harbored for many years? No; the eventual answer was far more idiosyncratic than that.

The workers' problems proved to be directly related to one particular work habit: licking their paint brushes to get a nice neat point. With each lick the workers ingested tiny bits of radioactive radium from the luminous paint on their brushes, and over time they had ingested huge doses of radium.[10] Radium, which is chemically similar to calcium, seeks out the calcium in bone and takes up residence there. Thus, the women's jaw bones disintegrated, in a process that came to be known as radium necrosis. Even worse, within the next 20 years, these workers started developing bone cancer at an alarming rate. At least 10 percent ended up dying from osteogenic sarcoma, a disfiguring and otherwise rare form of cancer.

In the 1940s and 1950s, as the atomic age dawned, the experience of the radium dial workers seemed especially relevant. Workers in the nuclear industry were exposed to radiation through uranium mining and from working with radioactive isotopes. Ordinary citizens were exposed through weapons testing and the transportation and disposal of radioactive wastes (though both of these sources paled in comparison to the single biggest source of radiation exposure for most individuals, medical x-rays). By about 1955, radiation moved from being an occupational problem to being everyone's problem.[11]

With nuclear weapons testing becoming a familiar part of the Cold War, "radioactive fallout contaminated the whole world with the deposition of strontium-90 in the skeletons of every living person," according to Vilma Hunt of the Harvard School of Public Health.[12] Naturally, people wanted to know what the health effects of such exposure would be. One clue, wrote Hunt, could be found in radium dial painters who were still alive and another could be found in the bodies of dial painters who had died. In a gruesome twist on scientific experimentation, the remains of the deceased workers were exhumed from their burial places so that they could be examined for radioactivity, which was still being emitted from their bones. These subjects, both living and dead, were studied intensively, wrote Hunt, "to establish the level of radioactivity associated with cancer and other changes in the skeleton and soft tissues." Ultimately, investigations like these led scientists to establish maximum permissible standards for all workers exposed to radiation.

Other scientists used occupational studies as a jumping-off point for a new field of investigation within radiation biology: the health effects of the radiation that people are exposed to naturally. This "background" radiation comes from degradation of ultraviolet radiation from the sun, emissions from certain rocks, and other ambient conditions. Background radiation is orders of magnitude weaker than occupational exposure—something like three-tenths of a rem per year, compared to 10,000 rem per year for radium dial painters—so any health effects would be much more difficult to discern. But John Little, in the early 1960s a young postdoc at the Harvard School of Public Health fresh from his training as a physician and radiologist, leaped at the challenge. He wanted to see whether there was an association between bone cancer and high levels of background radiation, as would be measurable in bone samples from bone cancer patients.

To obtain cancerous bone samples, Little asked his friends at the Massachusetts General Hospital for help. As he recalls:

> I went down there with my old and rusted Plymouth one day to get these bone samples. They had me back the car into the receiving area—and they proceeded to load into the trunk a dozen legs![13]

67

This stunned Little, who had thought he would be getting a test-tube-sized sample of tissue from each patient, not an entire limb. The legs had been amputated as treatment for bone cancer—providing Little and his co-workers with much more bone than they needed, plus several unwanted pounds of additional muscle, flesh, and skin.

> So there I was in my $75 car with the rusted trunk, and I got onto Storrow Drive just scared to death that one of those legs would fall out [of] one of the holes in the trunk.[14]

John Little

The legs stayed put, and after Little and his colleagues extracted the small samples of bone they needed, they followed what they hoped would be proper protocol and disposed of them in a hospital incinerator. As for their research, they found no association between the rate of bone cancer and radiation exposure as detected in bones. These negative results suggested that variations in background radiation exposure were not sizeable enough to account for differences in who did and who did not get bone cancer.

More recently, the issue of background radiation emerged as a result of concern about the naturally occurring product of radium known as radon. In the 1980s officials from the Environmental Protection Agency (EPA) said that radon was the single biggest cause of lung cancer after cigarette smoking. Federal estimates were that exposure to the decay products of radon gas in people's homes, called radon daughters, was responsible for thousands of cancer deaths each year. To alleviate the problem, the agency encouraged homeowners to have their basements tested for radon emissions, especially if they lived along the high-radon region in the northeastern United States known as the Reading Prong, named after the town in Pennsylvania that is at its center. In northern New Jersey, the hot zone of the Reading Prong, state health officials estimated that as many as one in three homes had dangerously high levels of radon.

Because of the efforts of state and federal health officials, radon became a household word in the 1980s and 1990s, as did the term indoor air pollution. The EPA set a limit for airborne concentrations of radon in homes above which remedial action should be undertaken. In 1994 Congress passed a law that required a radon inspection—much like the termite inspection already required—in any home about to change owner-ship. But no one knew precisely how risky radon exposure was and within a few years the problem dropped from the national consciousness. As Joseph Brain of the Harvard School of Public Health recalls:

> The EPA was saying that 10,000 to 20,000 people a year were dying from indoor radon. I think most people now think that was a little bit high. But almost everybody still thinks it's a significant problem and probably causing thousands of deaths, and it still needs some attention.[15]

The difficulty came from trying to quantify exactly how big the prob-lem was and exactly how much attention it required. What we know about low-level radiation risks involves extrapolation from what is known about high-level risks—and extrapolation always presents problems. What is known about the health risks of radon comes largely from stud-ies from the 1940s of men who worked in uranium mines, where radon levels were high, ambient temperatures were low, and breathing condi-tions involved exposure to a lot of dust, other toxic chemicals, and to-bacco smoke. The studies showed that male uranium miners exposed

Joseph Brain

to high levels of underground radon, especially those who smoked, had significantly higher rates of lung cancer than did a comparable group of men who did not work in the mines. "How do we get from that to women and children and nonsmokers who are exposed to far lower levels but for long periods of time?" Brain asks. That is one of the central questions in environmental health.

On the scientific graph of dose versus response, one cannot always be sure of a straight-line relationship, particularly at low doses. Sometimes, no response occurs at all until a dose is quite high; at other times even a vanishingly small exposure could theoretically cause problems, negating the possibility of a "safe threshold" dose at the lowest end of the scale.

While the general consensus has been that the dose-response curve for radiation is essentially a straight line—that is, the higher the dose, the greater the health risk, even for doses that are almost zero—there is some dissent on this point. Since the mid-1980s, a small group of scientists has been investigating the possibility that a very small dose of radiation is actually good for you. These scientists are resurrecting an old-fashioned idea known as hormesis, a word coined in the 1940s to describe the beneficial effects of exposure to low levels of substances— like fluoride, some infectious agents, and certain vitamins—that are toxic at high levels. Just as a vaccine of an attenuated virus can keep you from getting sick from a real-world dose of virus, hormesis implies that a small dose of background radiation, like the kind people are exposed to through radon in their homes, inures them to larger assaults later on.[16]

The health effect of low doses of ionizing radiation—the type of radiation transmitted by nuclear power plants, radon, or certain occupational exposures—continues to be a source of much debate and analysis within the public health community. In 1990 the National Academy of Sciences issued the fifth in its series of reports from its Committee on the Biological Effects of Ionizing Radiation, known by the acronym BEIR, which concluded that its own prior reports regarding the risk of cancer following radiation exposure had underestimated the danger. The conclusions of the BEIR committee were based largely on studies of cancer death rates among people who survived the atomic bomb blasts in Hiroshima and Nagasaki in 1945. The more recent results from this crucial epidemiological study of more than 75,000 bomb survivors led the BEIR V committee to estimate a risk that was three to four times higher than in its previous report, based on the data available 10 years earlier.[17]

It is hard enough for experts to predict the actual risks of low doses of radiation. Little believes the real risks, especially occupational ones, are very low compared to the risks of exposure to some chemicals that generate less fear because they are harder to measure. "People are frightened of radiation," he says, "because of these invisible waves that are going through us, which you can easily measure with a radiation counter." Indeed, the very ease of measurement is part of the problem in terms of perception. "You can detect radiation doses that are a thousand or ten thousand times lower than can produce any biological effect."[18]

An indication of the magnitude of the perceived risk was brought home years ago in a study comparing the opinions of a group of lay people—members of the League of Women Voters and college students—with those of scientific experts. As Little recalls:

> The investigators gave the lay people a list of 30 things we use in our lives and asked them to rank them in terms of risks. They all ranked radiation first—ahead of cigarette smoking, automobile accidents, and everything else. Then the same list was given to the panel of experts. They ranked radiation twentieth, right between food colorings and home appliances.[19]

Translating for public consumption the expert consensus on both real and perceived risks is a mission fraught with problems, which is perhaps why the public is so uncertain about how to view radiation risks. A case in point is related to radon. At the Harvard School of Public Health, Dade W. Moeller, who was trained as an engineer, helped develop a simple, inexpensive device that significantly reduced the concentration of radon decay products present in the air of average homes.

Confronting the Physical Environment

Dade Moeller

The success of our device was based on two facts, the first being that the primary hazard is not the radon gas itself but inhalation of a by-product of the gas as it decays, the solid radioactive products. The second fact is that the decay products are electrically charged. As a result, if the air within a room is gently circulated so that the decay products come into contact with surfaces in the room—furniture, table-tops—they'll adhere, exactly the way dirt accumulates on the screen of your television set, and be removed from the air. The net result is a reduction in the level of these products in the air by 50 percent, in other words, by a factor of two. Later, we found that if we combined ion generation with air circulation, we would reduce the amount of decay products in the air by 80 to 90 percent or even more.[20]

When Moeller and his associates tried to promote the application of their discoveries, a simple fan combined with an ion generator, they found that fear of radon had just about dropped off the popular radar screen. John Graham of the Harvard School of Public Health thinks this is a case of the public's misperception of risk. People are tired of hearing about all the things that might be hurting them and their children, says Graham, and are fatalistic about whether they can do anything to avoid radon in any event.

"The misperception of where the real risks are in this country is one of the major causes of what I call statistical murder," says Graham, who began the Center for Risk Analysis in 1990 as an outgrowth of his work with the Harvard Injury Control Center. "We're paranoid about Alar and nuclear waste sites in Nevada, and that preoccupation diverts attention from the real killers."[21]

Later in this chapter we will return to the notion of risk assessment and the perception of risk and will do so again in Chapter 5. For now it should be remembered that risk can take many forms: some visible, some not; some unusual, some encountered daily. In fact, one of the most common risks we face in the physical environment every day is the air we breathe.

Foul Air: The Deadly Smogs of Donora and London

It was the middle of the day in Donora, Pennsylvania, yet it looked as dark as midnight. On October 27, 1948, thick black clouds descended on the steel mining town's main street and the air smelled like rotten eggs. The weather reports on the radio talked about a "temperature inversion" causing the strange discolored fogginess. Old people and children were warned to stay indoors with the windows shut.

Donora is situated along the banks of the Monongahela River in western Pennsylvania, approximately 40 miles southeast of Pittsburgh. The air had always been foul in Donora; it was smoky and smelled bad, and it prevented grass or just about any other vegetation from growing. But no one complained, because in the 1940s nearly all of the town's

14,000 citizens depended for their livelihood, directly or indirectly, on the smoke-belching factories. Donora was a center of iron, steel, and zinc manufacturing, and the high-sulfur coal that ran the factories emitted noxious sulfur dioxide gas.[22]

During those few dark days in October 1948, though, Donora's air was more than just unpleasant; it was downright dangerous. And it killed more than vegetation. In retrospect, local physicians realized that visits to the town's emergency rooms had soared during the pollution alert. By the time the weather pattern shifted and the fog lifted, thousands of people had been hospitalized and 22 had died. Recalled one local physician about the four-day ordeal:

> I worked day and night. People were coughing and gasping for breath.
> In the worst cases, we had to use adrenaline, mostly on older people.
> But there was little we could do.[23]

Public health officials look back on Donora as the first clear demonstration that air pollution can cause illness and death. But in the generation since, the particular kind of air pollution that occurred in Donora has become so familiar that we have even given it a nickname. The toxic cloud that is part smoke and part fog is now known simply as smog.

Four years after the Donora episode, a similar smog descended on London. In early December 1952 the city—its air already fouled by the sulfur-emitting soft coal on which Londoners depended to heat their homes—was trapped beneath a temperature inversion. Soon there were the same black clouds as had occurred in Donora, the same thick fog, the same midday darkness, and the same deadly results. Because London was so much more populous than Donora, the number of deaths reached an astounding 4,000 before it was all over.

Douglas Dockery

This was a silent killer. If you read the London papers on December 5, you might have been told about the unusually dark fog but not a word about the toxic quality of the air or the deaths. "Nobody even knew what was going on at the time," says Douglas Dockery of the Harvard School of Public Health. As he puts it:

> It was just another foggy day in London. It was only retrospectively, when they went back and looked at the death records, that the statisticians saw a huge blip in the number of people who died.[24]

The keepers of London's vital statistics, combing through the prior year's records as they do on a regular basis, were stunned in 1953 to realize what had taken place the previous December. Once they saw the elevated death rates for the early part of the month, they began to search the records to see what was different about those days. And there it was in their laps: the deadly smog. It required no statistical sophistication to notice what was obvious at a glance: a remarkable correlation between elevated particulate and sulfuric oxide concentrations, which lasted for four or five days, and elevated death rates.

Most significant is that the death rates never went through a corresponding slump in the weeks after the pollution episode, which would have been expected if the poor air quality were only killing off those people who would have been expected to die anyway within a few more days. As Dockery explains,

> Immediately before the London fog, there were 250 deaths per day on average in London. During the days of extremely high particulate concentrations, there were 1,000 deaths per day. And after the episode the death rate was still elevated, at about 500 deaths per day.[25]

This was the first time that scientists could see a clear correlation between an increase in particulate concentrations and an increase in mortality. And because there was no corresponding decline once the air pollution abated, experts were convinced that the unusually high death rates represented a genuine increase in mortality, not merely a hastening of deaths likely to occur anyway.

London never again experienced as dramatic an episode of deadly air pollution. In the early 1950s city residents moved away from the use of soft coal for home heating fuel. But this change had nothing to do with environmental concerns; it was a simple matter of economics. At around this time, British workers discovered gas fields in the North Sea, and natural gas became cheaper than coal.

Shortly after the Donora and London episodes made it plain that air pollution had a direct effect on public health, the Harvard School of Public Health established its own division of environmental hygiene. The division was founded in 1958 under the direction of physiology professor James Whittenberger, a genial man of few words, whose carefully chosen comments were often taken as gospel by the dozens of protégés he trained during his 34 years at Harvard. Raised in a small town in Illinois—a town he liked to say was "so small that the bake shop had only one number"—he joined the Harvard faculty in 1946, at the age of 32, as an associate in physiology. Within a year he had become an assistant professor; within two years he was head of the department.[26]

Under Whittenberger's guidance, scientists in the division of environmental hygiene helped establish the connection between environmental exposure to certain airborne pollutants and subsequent increases in such problems as asthma, lung cancer, chronic respiratory

Confronting the Physical Environment

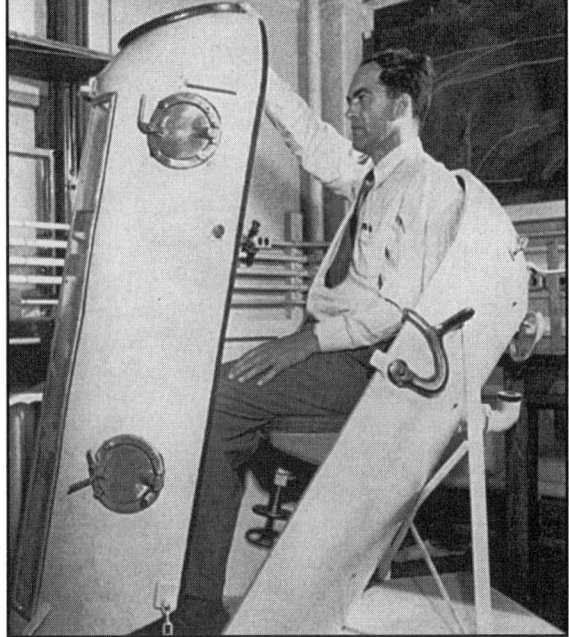

James Whittenberger with his Sit-Up Respirator

disease, and overall death rates. Many of these findings served as the scientific underpinnings for regulations enacted by the Environmental Protection Agency, which was established by an act of Congress in 1970.

The Clean Air Act of 1970, which created the EPA, focused on reducing emissions from power plants and automobiles. Ironically, though, the very steps taken to meet those standards might have created an air pollution problem even worse than the one they were designed to avoid. It wasn't until nearly 20 years later that scientists were able fully to understand the true nature of this paradox.

The most efficient ways to reduce emissions—installing cleaning devices such as scrubbers and filters on smokestacks or burning different types of fuel oil with lower sulfur concentrations—are also the most expensive. Many manufacturers, when faced with new EPA regulations, therefore opted for what seemed an equally effective, and certainly cheaper, solution: use the same fuel but raise the smokestacks. This would get the first emissions up beyond the "mixing layer" in the upper atmosphere, possibly keeping them aloft indefinitely and certainly allowing for a dilution of the chemicals before they wafted back to the ground where people could inhale them and EPA monitors could measure them. But this "solution" had an unintended result, says Dockery:

> When you get the pollution up in the upper atmosphere from these stacks, it will be transported for long distances before it gets to the ground. This provides an opportunity for the pollutants to change through chemical reaction while they're up there in the atmosphere. The sulfur oxide emissions can oxidize into sulfuric acid, which gets transported for long distances, and when rain falls the sulfuric acid gets deposited on the ground.[27]

That sulfuric acid is deposited in the form of the notorious acid rain. But it is not the acidity itself that causes the health problems, although in the 1960s and 1970s that is how the conventional wisdom had it. Indeed, for years the acidic nature of the Donora and London fogs was taken as an explanation for the high death rates during those events. But now the most dangerous component of air pollution is recognized as being particulates, microscopic pieces of solid matter that hang suspended in the air.

Information about particulates comes in large part from a long-term investigation, known as the Six Cities Study, conducted at the Harvard School of Public Health. The study began in 1974 by Benjamin Ferris in response to the energy crisis, which was turning the nation more and more to coal as an energy source. Burning coal leads to far more sulfur dioxide in the air than does burning oil, and the move back toward coal had many experts worried about its effect on respiratory health.

For 12 years Ferris and his colleagues traveled to six cities—low, medium, and high pollution cities—to record the respiratory health status and pulmonary function of about 12,000 children who were in first grade when the study started, as well as nearly that many adults. They also measured sulfur dioxide, suspended particles, nitrogen oxide, and ozone at selected indoor and outdoor sites. They found that deaths from lung cancer, lung disease, and heart disease were 26 percent higher in the most polluted of the six cities (Steubenville, Ohio) than in the least polluted one (Portage, Wisconsin). Yet even the dangerous air in Steubenville still meets federal standards for particulates.[28]

In 1986 a strike by steel workers at a plant in the Utah Valley provided a dramatic natural experiment to test the relationship between particulate pollutants and human health. When the plant—an old World War II-era factory with poor emission controls—was in operation, cold winter air and pollutants were trapped in the valley and the air was filled with sulfur dioxide and other toxins. When strikers shut down the plant, the air quality improved dramatically, with particulate concentrations just about half of what they had been before. The strike lasted 13 months, and, though families were strained economically, the air they breathed was far healthier than what they were accustomed to. In the fall and winter immediately before the strike, 127 children from the Utah Valley were hospitalized for bronchitis and asthma and another 96 for pneumonia and pleurisy. In the year when the strike was on, the number of bronchitis and asthma admissions dropped by more than 60 percent to 46, and pneumonia and pleurisy admissions dropped to 57. In the year following the strike, admissions rates were up to exactly what they were before the plant closed.[29] As Dockery observes:

> It's very unusual that you have an environmental intervention like this, turning off the source of pollution and turning it back on again. This was a landmark study, allowing us to look clearly at the effects of air pollution.[30]

Dockery believes that, based on what is now known about particulates, the standards outlined in the Clean Air Act are set too low. He says particulates in the air work much the way indoor dust works on radon decay products. They efficiently transport gaseous pollutants deep into the lungs, evading the tiny hairs and mucous membranes of the nose and mouth that can trap and remove most nonparticulate toxins. Once inside the lungs, where no natural barrier exists to screen out dangerous chemicals, pollutants have the potential to destroy living tissue. Estimates are that in the United States there are 20,000 to 30,000 premature deaths each year from respiratory failure that can be directly traced to particulates.[31]

Difficult as the task of setting standards is, at least they do exist in the United States. Most developing countries have no such governmental oversight. And because their factories and cars are old and lacking in emission controls, their cheap heating sources (wood and animal dung) are unusually dirty, and their gasoline is almost always leaded, the airborne particulate readings in these countries are usually at least 10 times higher than the EPA would allow.[32]

Another, even more pervasive, form of pollution—water pollution—is also a devastating problem in the developing world. Questions of water purity go back to concerns that in the United States were handled more than a century ago—water sullied with human wastes and bacterial contamination. Many villagers in underdeveloped nations still excrete directly into the streams where they get water for cleaning clothes, cooking dinner, and quenching the thirst of their animals and themselves. As one journalist reported recently, 1 billion people worldwide rely on water supplies that cannot meet even the crudest safety standards.

> Most of Africa, the Indian subcontinent and Latin America have no waste-water treatment facilities; raw human and industrial sewage is discharged directly into the same bodies of water used for drinking. Mothers draw for their children water of lower quality than the process effluent discharged from American factories.[33]

As a result of such crude water supplies, nearly 4 million children under the age of 5 living in underdeveloped countries died of diarrheal diseases in 1993—about the same as the total number of people of all ages who died of all causes in all of Europe and the United States.[34] So while simple water purification seems to be taken for granted by the American public, international public health efforts are still very much involved in getting flushing toilets and sewage systems to the megacities and villages of the developing world.

Water purity is still an issue in the Western world as well, but it has taken a different form. In the United States, as we shall see, a threat to our drinking water involves the presence of toxic chemicals. The precise nature of the threat has been the subject of an ongoing debate among scientists, politicians, and industrial polluters, and we turn now to a focus of a large part of this debate: two of the wells supplying water to a single town in northeastern Massachusetts.

Foul Water: The Poisoning of Woburn

In the 1940s almost any problem with water purity could be cleaned up by giving a town enough flushing toilets. The biggest hazard to public health was human waste, which in overburdened sewage systems—found in both squalid urban slums and otherwise bucolic country settings—allowed the commingling of wastewater and drinking water. This was the prime breeding ground for water-borne diseases such as cholera, which remains a major health threat in many parts of the world but had all but disappeared from most parts of the United States by the early 1900s. As public health historian George Rosen wrote:

Those great epidemic terrors—cholera and yellow fever—disappeared, never to return, before the specific causes of these infections were discovered and before exact knowledge of their transmission became known. These trends undoubtedly reflect in part the impact of the earlier sanitary reform movement. Acting on the theory that "a clean city is a healthy city," housing was improved, the physical environment was cleaned up, efforts were made to provide unadulterated food and clean water.[35]

But today, despite nearly a century of impressive sanitary improvements, America's drinking water is still not perfectly safe. The difference is that not only do we have to worry about microbial contamination[36] but also contamination by industrial chemicals.

That is what happened in the small industrial town of Woburn, Massachusetts, about a half-hour's drive from downtown Boston. In 1972 Anne Anderson, a mother of three from Woburn, brought her youngest child, three-year-old Jimmy, to the Harvard-affiliated Children's Hospital for treatment for leukemia. While spending time waiting in the leukemia ward, "she noticed a lot of familiar faces around, and they were from Woburn," recalls Marvin Zelen, a Harvard School of Public Health biostatistician who became involved in the case. By 1976 Anderson had counted a total of six cases of childhood leukemia in Woburn—enough to convince her minister, Reverend Bruce Young, that something suspicious was afoot.

At about the same time, officials of the Massachusetts Department of Environmental Quality and Engineering were amassing some information of their own. Construction workers had uncovered several barrels of waste buried near two of the eight wells supplying Woburn's water. When the wells were tested for toxic pollutants, the water showed high levels of trichloroethylene, tetrachloroethylene, and chloroform, all of which are known to be animal carcinogens. The wells were closed. Later, after large pits of buried animal hides and chemical waste were uncovered in nearby marshes, the groundwater that fed those same two wells was found to contain 48 dangerous pollutants and 22 toxic metals. As journalist Douglas Hand later put it:

> One might think that bits of information would then have come crashing into place, like tumblers in a lock, and some public official would ask, at the least, "Is something wrong here?" Instead, Young and Anderson still faced what Young calls the "universal script": A mother says there's a problem. Health officials deny it. Eventually, they demand, "You do the work and prove it."[37]

Zelen remembers when he decided to "do the work and prove it." Young and Anderson came to speak at a seminar at the Harvard School of Public Health in 1980, when Zelen was chairman of the Department of Biostatistics. They brought with them a town map on which they had plotted the homes of 12 Woburn children who had leukemia; half of them lived near Anderson's home. They reported that the Centers for Disease Control had been alerted and that CDC scientists, in collaboration with the Massachusetts Department of Public Health, had confirmed their worst fears: the incidence of childhood leukemia in Woburn was more than twice the expected rate; within half a mile of Anderson's home it was 7.5 times higher. The federal report, though, made no conclusions about the possible causes.

Marvin Zelen

Anne Anderson must have seemed a tragic figure at that Harvard seminar. Only a few months earlier her son Jimmy had died; he was 12 years old and had battled leukemia for nine years. Rev. Young came with her to offer his own perspective to the students and faculty members. He said his goal was to tell them, "Hey, not every mother who comes in claiming there's a problem is a kook."[38]

The audience, including Zelen, was impressed. Zelen and his associate, Stephen Lagakos, got right to work:

> Steve and I decided to carry out a study. We asked Anne and Bruce if volunteers could be recruited to help carry out a study. They felt it was possible. We enlisted about 300 volunteers, trained them in a workshop, and put them on the telephone to get data from the residents of Woburn about not only leukemia, but birth defects and other childhood illnesses.[39]

Within a few months, Zelen and Lagakos had collected data from about 7,000 town residents. An association was found between proximity to the two contaminated wells and a higher rate of leukemia as well as some birth defects.

The scientists informed Anderson and Young about what they had found, and they asked them to convene a meeting with the community the following evening. Young arranged it at his church, Trinity Episcopal. The next evening, Zelen recalls:

> Steve, myself, my wife, and an assistant were journeying to Woburn, and we thought there would be about 25 people there. We hadn't prepared very much, a few hand-drawn transparencies. We thought we'd sit around and just explain what we found. We got into the church parking lot, though, and we knew something was up. We couldn't find a parking place.[40]

What the Harvard contingent found was a huge crowd: hundreds of Woburn residents, national press, television crews. Zelen says his part in it was a "disaster—academic double-talk." He was not yet familiar with being in the limelight.

But the limelight was inevitable because of the kind of study that Zelen and Lagakos had prepared. *The New York Times* hailed it as "a major breakthrough in the field," considered by many experts to be "the first in this country ever to correlate the distribution of contaminated water with an outbreak of childhood leukemia."[41] What made the work especially noteworthy was that the scientists had demonstrated a dose-response relationship between contaminated water and disease. The more poisoned water a prospective pregnant mother was exposed to, the more likely her child was to have cleft palate, Down syndrome, or other anomalies. The more contaminated water a child was exposed to, the greater the likelihood of leukemia.

"The findings of the Harvard researchers were greeted with a sense of relief among Woburn residents," wrote Jonathan King, an investigative reporter. "They felt they were proof that their drinking water was making them sick. But the Harvard study also added fuel to an already raging debate over how harmful the chemical contaminants in our drinking water really are."[42]

Despite the general acclaim that greeted the Harvard study, it received more than its share of criticism. Some critics said Zelen and Lagakos had worked too closely with community activists and depended too much on community volunteers for data collection to allow for a truly objective study. According to Phil Brown, a medical sociologist at Brown University who reviewed the controversy, charges of study bias came from a variety of groups, including the Massachusetts Department of Public Health, the Centers for Disease Control, the American Cancer Society, the EPA, and even the Harvard School of Public Health's Department of Epidemiology.[43]

Some of the criticism came from within the Harvard School of Public Health itself, most from Brian MacMahon, who at the time was chairman of the Department of Epidemiology. In the same issue of the *Journal of the American Statistical Association* in which Zelen and Lagakos's study first appeared, MacMahon published a rebuttal in which he made the following conclusion:

> The investigators have brought forward some associations that may warrant further investigation in other geographic areas, but to state that there was "a consistent and recurring pattern of positive associations [of childhood leukemia and birth defects] with availability of water from wells G and H" as the authors have in a previous publication is to grossly overinterpret the data. [They] have moderated their interpretation (in the right direction), but there remains a way to go to where I stand.[44]

In retrospect, Steve Lagakos views the academic debate—which was always polite but nonetheless deeply felt—as a matter of differences in basic political philosophy. As he recalls:

> A lot of it came down to, I felt, the issue of burden of proof. If the attitude is to assume that these chemicals cause no harm unless it can be proven unequivocally that they do, I don't think any epidemiological study, and certainly not ours, could prove anything. On the other hand, if the attitude is that the burden of proof is on the other side to show that the chemicals didn't cause harm, you have a totally different perspective. I think the industry's attitude, and certainly Brian MacMahon's, falls into the first perspective. The residents of Woburn by and large believed in the second perspective.[45]

After the Harvard study was first published in 1984, seven Woburn residents filed suit against W. R. Grace Chemical Corporation and Beatrice Foods, accusing them of improperly disposing of waste products from their facilities and endangering the community's health. The lawsuits, which were the subject of the nonfiction bestseller *A Civil Action*, were eventually settled out of court.[46]

When the residents of Woburn believed their water had been poisoned, they sprang into action in part because of the powerful metaphor of water itself. Water is crucial for life; no plant or animal can survive for long without it. It is clear, refreshing, and seemingly pure to the naked eye. Maybe that is why people feel so betrayed when their drinking water is tainted; the danger is invisible, tasteless, and pernicious. And maybe that is why—even before the real threat to drinking water in communities like Woburn and Love Canal—there was such outrage expressed in the 1950s when public health officials wanted to add fluoride, which they promised was harmless, to the drinking water of many American cities and towns. The notion that government authorities might tinker with a community's water supply turned out to feed into some rather dramatic fantasies that captured the paranoid emotions of the Cold War at its height.

Was Fluoridation a Communist Plot?

In the 1950s the addition of fluoride to drinking water was promoted as an important way to reduce cavities. It is hard to remember how big a problem cavities used to be. Today, a typical American child grows up without getting any at all. This improvement is due in part to dental care innovations like long-term topical fluoride treatment and protective plastic sealants on the permanent molars, and in larger part to public health innovations such as fluoridation of water supplies. Kids today have strong teeth, and the majority of them—about 65 percent of 9 year olds, according to a recent study—have no cavities at all in their permanent teeth.[47]

Just a generation or two ago cavities had an enormous public health impact. Bad teeth were one of the most common reasons for rejection by the Selective Service System during World War II.[48] About one in 10 of the 4 million young men examined was rejected for military service because he lacked the six opposing teeth required to "bite the cartridge."[49]

Tooth decay was so common that dental epidemiologists would calculate oral health by going into a community and asking people to open their mouths for a DMF count—the number of teeth that were "decayed, missing, or filled." Once cavities started eroding the teeth, they could cause great pain and loss of time at work. If the teeth had to be removed, or fell out on their own, the resulting gap could cause new problems in chewing, speaking, or even breathing.

The difference between a mouth full of cavities and a mouth full of strong, healthy teeth is a compound called fluoride. The link between fluoride and dental health was first established not because of how fluoride helped the teeth but because of how it hurt them. In 1902 Dr. Frederick C. McKay, a young dentist from Colorado Springs, noticed that many of his patients had discolored teeth, a kind of weird brown freckling, or mottling, that marred the enamel. He wondered why the discoloration was happening and why it seemed to happen in some towns scattered across the country but not in others. "Something in the water," was the folk wisdom about those mottled teeth. But chemical analysis at the time was crude, and fluorine (the main component of fluoride) is an active element with a short life span. No tests were sensitive enough to detect it.

McKay was especially intrigued because he had noticed that his patients with mottling were also those whose teeth were the sturdiest. In 1908, and again in 1925 and 1926, he published reports showing that mottled teeth had very low rates of decay. It seemed that whatever it was that was so bad for the tooth's appearance was good for its health.

In 1923 McKay convinced the townspeople of Oakley, Idaho, to put the "something in the water" theory to the test. At the time, just a few years after the town's fathers changed the source of their drinking water, mottling had begun to appear in the teeth of Oakley's youngsters. McKay managed to convince them to change the water supply again to a third source, at a cost to the taxpayers of some $35,000. Luckily, the third water source had a lower fluoride content than the previous one and McKay's theory was confirmed. Some Oakley children had mottled teeth, but children who were just a little older or just a little younger did not. The difference did indeed seem to be "something in the water."

In 1930 an experiment conducted by the chief chemist at the Aluminum Company of America (Alcoa) identified that critical "something." The Alcoa chemist was trying to prove that the mystery ingredient was not aluminum, which was under suspicion because of a high rate of tooth mottling in the aluminum ore-mining town of Bauxite,

Arkansas. McKay sent the chemist samples of Bauxite water, and the Alcoa chemists,

> by running a very complete analysis, found a large amount of fluoride. They then analyzed the water from other mottling-prone towns, and found the fluorine ion in every case. McKay's problem was finally solved.[50]

Soon afterward, these findings were confirmed by animal experiments: a husband-and-wife team in Arizona showed that rats who drank fluoridated water had mottled teeth—which, as scientists already knew, were sturdier teeth. At the time, another young dentist, H. Trendley Dean, was also hot on the trail of fluoride, traveling from town to town to chart the degree of mottling, the rate of tooth decay, and the level of fluoride in the water supply. In 1941 he came up with a formula: one part per million of fluoride in the water was high enough to reduce cavities and low enough to avoid staining.

Three years later the U.S. Public Health Service began a large-scale, 10-year fluoridation experiment involving four mid-sized cities. Grand Rapids, Michigan, and Newburgh, New York—the experimental towns—would receive one part per million of fluoride in their drinking water. Muskegon, Michigan, and Brantford, Ontario—the control towns—would not. Within three years it was clear that fluoridation had made a difference. Children in Newburgh, for instance, experienced a one-third drop in dental decay.

The final experimental results were expected in 1954, but even before that time health officials in many towns tried to fluoridate their own water supplies. Why waste time waiting for the trial to be over when the early evidence was so compelling? As one early advocate explained, it would not be fair to "write off one generation while they see if two and two makes four."[51] Public health officials at the time lacked the statistical sophistication to understand the need for large-scale, longitudinal, controlled clinical trials. If they had waited, the scientific case for fluoride might have been made even more forcefully—and might more effectively have withstood the assaults on fluoride from nonscientific quarters.

Convincing the population at large of the value of fluoride was much harder than convincing the public health community. When voters were asked to approve, by ballot, the fluoridation of their own water supplies to protect their children's teeth, many of them flatly said *no.* Opposition to fluoridation was loud and strident. Some critics charged that fluoridation was "another aspect of President Truman's drive to socialize medicine."[52]

A common scene was this one from Seattle, where a fluoridation referendum was held in 1952:

> The opponents included an organization of chiropractors and homeopathic physicians who painted a hideous picture of the physical effects of fluorides.... A group of 20 dentists with a strong interest in natural

foods joined the battle; they opposed fluoridation largely on nutritional grounds. One of these, Dr. A. B. McWhinney, became an outspoken advocate of other methods of distributing fluorides; incidentally, he owned a company which manufactured fluoride tablets. . . . [Still other opponents] concluded, loudly and often, that fluoridation was a Soviet conspiracy to annihilate Seattle (and Boeing's bomber plant) with a slow poison. Finally, a number of Christian Scientists argued against fluoridation as compulsory medication.[53]

Fluoride referenda were voted down in more than half of the 2,000 communities that considered them in the 1950s and 1960s. Opponents were rabid in their belief—and their organized dissemination of that belief—that water fluoridation was a Communist plot to deplete the brainpower and sap the strength of a generation of American children.

Most of the opponents, as Edward O'Rourke noted in 1953, were "food faddists, members of religious groups opposed to medicine in general, and those self-appointed guardians of the public who appear before most open committees." At the time, O'Rourke, a graduate of the Harvard School of Public Health, was the health officer of the city of Cambridge, Massachusetts, where open debates about the pros and cons of fluoridation were loud and lively. One of his biggest frustrations in watching the deliberative process in action was that

> equal time is given to the opposition in accordance with democratic procedure. Thus, one dentist who represents all members of the Massachusetts Dental Society and their considered evaluations is answered by a dentist who represents no one but himself. This gives the general public the erroneous idea that professionals are about evenly divided.[54]

Despite the struggles to get fluoride referenda passed in the 1960s, the public health community finally won out and the nation's children were the beneficiaries. By the early 1990s, virtually all American communities either had naturally high fluoride levels in the water or had added fluoride to the drinking supply to achieve the one-part-per-million standard.

The American fluoridation saga captures the struggle in public health to balance benefits to the public against risks (either real or, in this case, perceived) to individuals. At what point are public health officials justified in intervening on a community-wide basis to protect a group of people who are not all equally at risk and who might not want to be protected? The push and pull of paternalism versus autonomy is a constant refrain in the field, having been raised in relation to many familiar public health quandaries: whether to iodize salt, whether to add folic acid to bread, whether to require people to buckle their seat belts or to wear bicycle helmets, whether to limit the places where they can smoke cigarettes. One notorious example of the paternalism of public health measures—something against which people of all political stripes have found themselves rebelling ever since it appeared in 1958—is a federal law known as the Delaney clause.

The Delaney Clause and the Quest for a Risk-Free World

In the 1950s and 1960s traces of poisonous chemicals seemed to be everywhere: in the water we drank, in the tunafish sandwiches we fed our children, in the shampoos we used, and in the very air we breathed. The concern culminated in the 1962 publication of Rachel Carson's bestseller, *The Silent Spring*, which led Americans to take a new and skeptical look at a whole range of synthetic chemicals that threatened to undermine the health and well-being of industrialized societies. "It is simply impossible to predict," she wrote, "the effects of lifetime exposure to chemical and physical agents which are not part of the biological experience of man."[55]

It was in this atmosphere that the Delaney clause was passed. In 1958 James Delaney, a Democratic congressman from New York, inserted an amendment into the Food, Drug, and Cosmetic Act stating that any compound "found to induce cancer when ingested by man or animal" must be banned from the marketplace. That ban applied no matter how low a level was found in the food or drug involved and no matter how high a level was required to cause cancer in the experimental animal.

In 1958 the language of the Delaney clause was extreme but acceptable. With the measurements available at the time, any food additive or pesticide residue that existed in quantities high enough to be detected probably should have been subject to such rigid scrutiny. But in the years since, with the Delaney clause still on the books, science has progressed to the point where compounds can be detected in foods at trace levels that are vanishingly small. With investigators now trying to prove the safety of additives that exist on the order of parts per billion—several orders of magnitude smaller than what could be detected in the 1950s—the whole exercise begins to look a little absurd.

"The Delaney amendment has been basically ignored for 15 or 20 years, particularly as it applies to pesticide residues in processed foods, because it's not practical," says John Graham of the Harvard School of Public Health. "It was well intended but misguided."[56]

Among the early victims of the Delaney clause was the artificial sweetener cyclamate. In 1969 laboratory experiments showed that cyclamate, the most popular artificial sweetener on the market (sold as a liquid called Sucaryl), could cause cancer. The tests were the standard toxicological tests of the day for screening chemical compounds for possible human carcinogens. One, for instance, involved giving very high doses of cyclamate for a specified period of time until one-half the animals died from direct toxic effects of the chemical. This dose, known as the Lethal Dose-50 (LD-50, for 50 percent), was usually far higher proportionately than any dose a human would ever receive. But at these highest levels enough animals sickened or died to allow scientists to examine the toxic effects of the chemical without using too many animals.

When the LD-50 was reached, the toxicologist killed the rest of the experimental animals and conducted autopsies to see which organs were damaged. Scientists looked in particular for signs of cancer. In the case of cyclamate, a significant proportion of the rats and mice exposed to an LD-50 dose of the sweetener developed cancer of the bladder and other organs. That finding was enough, under the Delaney clause, to take Sucaryl off the market.

Critics of the cyclamate studies made much of the fact that, for a human being to be exposed to as much cyclamate as the rodents were, he or she would have to drink dozens of cans of diet soda every day for a lifetime. But the Delaney clause was clear on the matter: cyclamate caused cancer in experimental animals, and therefore it had to be banned.

There things might have stayed had not another artificial sweetener, saccharin, faced a similar ban just a few years later. In 1977 a report from Canada confirmed what other studies had found: saccharin caused bladder cancer in laboratory animals. But if saccharin should be banned, there would be no artificial sweetener on the market at all. As a *Consumer Reports* article summarizing the controversy in 1985 put it:

> A public outcry, eagerly fanned by the soft-drink industry, ensued.
> Congress, listening to the protests, imposed a moratorium on the sac-
> charin ban. It has extended the moratorium every few years since.[57]

Subsequent studies seemed to indicate that saccharin is a more potent carcinogen than originally presumed and that cyclamate may be even less potent. Regulators then found themselves in the position of allowing a more dangerous carcinogen to stay on the market because they had already banned the less dangerous one, though neither was a particularly large risk in the overall scheme of things.

The inequities and inconsistencies that have grown out of the Delaney clause have concerned legislators for years. But they do not quite know what to offer to replace it. If they propose an easing of the Delaney requirements, they might find themselves in the politically untenable position of voting in favor of suspected carcinogens. The problem is that some of the chemicals under scrutiny are found in food-stuffs at such low concentrations that, even if there is a chance they cause cancer in animals, they are unlikely to create problems in humans. "This gets back to the problem of safe and unsafe," says John Graham, who is making a formal study of the relative magnitude of certain public health problems. All that people know, he says, is "gee, it causes cancer, and it's in my food." But he wants them to consider the alternatives. Take fungicides, for example:

> People don't realize the kind of contribution fungicides make in re-
> ducing the cost of bringing fruits and vegetables to the grocery store.
> Now, there's pretty good evidence that fruits and vegetables reduce
> both heart disease and cancer. If you had even a modest increase in
> the cost of fruits and vegetables as a consequence of not allowing fun-

gicides to be used, you could be involved in causing more heart disease and cancer—particularly in the low-income population, where you don't want to do anything to discourage fruit and vegetable consumption."[58]

One important limitation of the Delaney clause is that it is restricted to food and drug components that have been shown to cause cancer. But some of them cause many other problems besides cancer or cause cancer that does not show up in people until decades after they are first exposed. That is what happened in the case of a widely prescribed hormone known as DES, which reached its heyday in the very same years that the Delaney clause was under debate. With the best of intentions, thousands of American women and their doctors found themselves caught in a relentless nightmare—a tale that can serve as a parable of the unintended consequences of medical intervention, consequences that often do not come to light until they are scrutinized by some dedicated individuals who are alerted to an emerging problem through the tools of epidemiology.

DES: In Our Medicine and in Our Food

Having babies was practically a national pastime in the 1950s and 1960s: more than 3.5 million babies were born every year between 1947 and 1953 and more than 4 million each year from 1954 to 1964.[59] And as pregnancy became more commonplace, it also became more medicalized. The Baby Boom was characterized by an unprecedented use of drugs during pregnancy. Between 1958 and 1965, according to one study, half of all new mothers took two to four pharmaceutical products during their pregnancies. These included aspirin, tranquilizers, antibiotics, and hormones. Physicians believed, as did most Americans, that modern scientific know-how could be successfully applied to pregnancy as to many other physiological disruptions and that drugs could help avert anything from pregnancy's small discomforts, like morning sickness, to its major disasters, like miscarriage. "Taking drugs was a part of the national belief system," writes journalist Robert Meyers about this period. "If America had won two wars overseas, it could win the war on health problems, including miscarriages."[60]

Among the drugs liberally dispensed to pregnant women was the artificial estrogen known as DES (short for diethylstilbestrol). During the Baby Boom years, DES prescriptions during pregnancy were commonplace, usually to prevent miscarriage or bleeding. Some obstetricians prescribed the drug almost routinely, even to low-risk women, just in case problems arose. By 1957 one manufacturer was advertising DES "for routine prophylaxis in ALL pregnancies . . . bigger and stronger babies, too."[61] But the dark side of this massive drugging of the unborn slowly emerged in the late 1960s. That is when a medical convention—

using hormones to grow bigger, better babies—translated into a public health phenomenon: the development in these babies of disorders of the genitals and reproductive tract, including infertility, abnormal pregnancies, and cancer.

The DES story began to unfold on the Fourth of July 1967, when a 16-year-old girl was admitted to Boston's Massachusetts General Hospital so physicians could investigate the cause of her vaginal bleeding. The news was not good. After careful examination and biopsies, the chief of gynecology at the Harvard-affiliated hospital, Howard Ulfelder, diagnosed a rare form of vaginal cancer called clear-cell adenocarcinoma. Within three weeks Ulfelder had removed the teenager's uterus and vagina and constructed an artificial vagina using skin grafts.

The cancer and its treatment were disturbing enough, but what especially worried Ulfelder was an apparent trend. Clear-cell adenocarcinoma had previously been seen primarily in postmenopausal women and never before in teenagers. Yet this 16 year old was the third young woman he had seen in three years with this rare cancer. "There must be some explanation for this explosion," the doctor said to his nurse.[62]

In early 1969 Ulfelder met again with the girl and her mother for a follow-up visit. "I don't think I ever told you something," the mother said. "When I was pregnant with her, the doctor put me on stilbestrol [DES] because I had lost one pregnancy. Could that have anything to do with it?"

This question raised Ulfelder's radar. Later, he asked the mother of another young vaginal cancer patient whether she had also taken stilbestrol. The woman said yes. And later still one of his colleagues at the Harvard Medical School, Arthur L. Herbst, asked the same question of the mother of one of the first teenagers seen with vaginal cancer. That girl's adenocarcinoma had appeared in 1964, when she was 14 years old; she died four years later. As Herbst remembers it, the mother told him, "You can stop your study. I took DES."[63]

Of course, this didn't mean the study should stop; indeed, it meant a serious study should now begin. If the administration of DES to pregnant women turned out to cause cancer in their adolescent daughters, this was a potential public health problem of major significance. Estimates were that in the 1950s and 1960s at least 2 million pregnant women took DES.

The public health implications of Ulfelder and Herbst's findings stimulated the two physicians to walk across the courtyard of the Harvard Medical School campus on Longwood Avenue to consult with their associates on Shattuck Street, at the Harvard School of Public Health. A biostatistician at the school, Theodore Colton, helped the physicians design a study that would reliably establish a relationship, if one existed, between intrauterine exposure to DES and vaginal cancer. The statistics involved were tricky because the cancer was so rare. Indeed, the eight cases that Herbst and Ulfelder found of clear-cell adeno-carcinoma in women younger than 22 were nearly equal to the number of previous reports of this form of vaginal cancer in the entire world literature.[64] The scientists were wary of designing the study as though they already knew the outcome and consequently overlook some unanticipated correlation. "When you design a questionnaire," explains Herbst, who is now chief of obstetrics and gynecology at the University of Chicago's Pritzker School of Medicine, "you've got to be careful not to design it in a way to provoke the response you may be looking for."[65]

The study design was simple and elegant. For each of their eight cancer cases the investigators chose four controls: women who were born within five days of the cancer cases, in the same type of hospital setting (public versus private), but who differed from the cases in that they did not have vaginal cancer. They then devised a lengthy questionnaire for the 40 young women. Regarding the mother's pregnancy, they wanted to know the mother's age when she delivered, her smoking habits, whether there was bleeding during her pregnancy, prior pregnancy loss, intrauterine x-rays, whether the mother breastfed, medications used during pregnancy, and of course whether the mother took any forms of estrogen while pregnant. Regarding the young women themselves, the researchers asked about birth weight, age when menstruation began, use of douches or tampons, childhood diseases, history of tonsillectomy or other operations, history of birth control pill use, smoking history, use of cosmetics, and presence of household pets—a wide range of questions that reflected the open-ended nature of the inquiry.

The results were dramatic. Of the eight mothers whose daughters later developed vaginal cancer, seven had taken DES during pregnancy; all of them began taking it during the first three months and all because of bleeding problems. Of the 32 control mothers, not one had taken DES.

The report of their findings appeared in *The New England Journal of Medicine* in April 1971.[66] Four months later the journal published a corroborating article by Peter Greenwald, director of the Cancer Control Bureau of the New York State Department of Health. Greenwald had checked the state's cancer registry for cases of clear-cell adenocarcinoma and found five. All of them were in women born between 1951 and 1953, and all of the young women's mothers had taken DES when they were pregnant with them.[67]

By the end of 1971 the scientific community was confident that there was a cause-and-effect relationship between DES exposure in utero and clear-cell adenocarcinoma in adolescence. And data were accumulating about other hazards from DES as well. For the young women whose mothers had taken the hormone during pregnancy, there were more disorders of the reproductive tract: abnormal cell growth in the vagina and cervix; structural irregularities of the cervix and uterus; infertility; and a higher risk of ectopic pregnancy, miscarriage, stillbirth, and premature delivery. The mothers themselves were shown to have a 50 percent higher risk of breast cancer. And in 1979 William J. Dieckmann and his colleagues at the University of Chicago published a study that indicated that the hormone did not spare males. Dieckmann's study, comparing 308 DES-exposed young men with 307 age-matched controls, found that DES sons had four times the rate of hypoplasia of the testes, 17 times the rate of undescended testicles, and four times the rate of microphallus (small penis).[68] Subsequent studies found not only structural abnormalities but also fertility problems caused by lowered sperm counts or malformed sperm.[69]

Once the relationship between DES and subsequent problems in the fetus became clear, the Food and Drug Administration (FDA) announced that DES was contraindicated for use in pregnant women. Virtually no DES has been prescribed for this purpose since 1972. But obstetrical use turned out to be a secondary factor in terms of how most Americans were exposed to the artificial hormone. The main source has been through DES residues in meat and poultry. Beginning in the 1940s, farmers learned that mixing DES into animal feed, or inserting it in the form of hormone pellets into the necks of chickens or cattle, made animals get fatter while eating less. Thus, DES was seen as a way to significantly increase the profits of the food industry. By 1970 nearly 75 percent of all cattle in the United States were being given DES.[70] But all those hormone-fed animals contained estrogen residues that could easily end up on the dinner plate, exposing millions of Americans, including children, to artificial hormones. No clear association has ever been made between exposure to hormones in meat and a specific health risk, but public health officials are suspicious that such a link might exist nonetheless. In keeping with the Delaney clause, FDA regulations required that the meat or poultry that made it to the marketplace be free of all DES residue, since it had been shown to cause cancer in laboratory animals.

In 1959 the FDA banned DES in chicken feed.[71] Cattle ranchers were exempted, and three years later Congress formally allowed ranchers to mix DES in their livestock feed provided they withdrew it at least 48 hours before slaughter, theoretically to remove all traces of hormones in meat that got to the grocery store. This regulation proved difficult to enforce, though; of the tiny fraction of animal carcasses tested by the FDA (1,000 a year out of the 30 million slaughtered annually), residues of DES were regularly detected.[72] By 1972 the waiting period between DES withdrawal and slaughter was extended to seven days—but still the amount of residue found in meat quadrupled from the previous year.[73] The National Cancer Institute joined forces with several members of Congress to urge a total ban on DES in animal feed, at least pending the outcome of the public hearings to be sponsored by the FDA. Bowing to pressure from politicians, scientists, and the requirements of the Delaney clause, FDA Commissioner Charles Edwards announced a ban on DES feed additives in August 1972. He did so grudgingly and made no secret of his feelings. "Under the law," he said, "there is no alternative but to withdraw approval of the drug, even though there is no known public health hazard resulting from its use."[74]

Some observers have questioned whether the ban on DES in meat was a Pyrrhic victory, given the fact that a wide range of similar estrogen-like compounds are still being used in livestock. Once publicity over DES in animal feed died down, notes Diana Dutton, a health policy analyst at Stanford University's School of Medicine, the FDA "quietly eliminated" the 60-day withdrawal period originally required for Synovex, another artificial estrogen used as a growth hormone. In addition, writes Dutton, "antibiotics, pesticides, fungicides, and other toxic chemicals are routinely used in food production."[75]

Even if potential carcinogens or other toxins were to be eliminated from the food supply, one problem still remains: environmental exposure to some of the very same chemicals. For some of them, no level of exposure, no matter how minuscule, has been proved to be safe. The most significant is one against which great strides have been made in this country in the past 25 years but one that is still probably injuring thousands of children. This toxin, lead, has been blamed for everything from the fall of the Roman Empire to a lag in IQ scores among inner-city children—and its effects permeate our country even today.

The Legacy of Lead

The Australian state of Queensland was barely settled in the 1890s when its children started dying horribly. Some 200 children a year, a significant proportion of the young population, suffered from a syndrome that began as headaches and nausea and soon progressed to cramps, seizures, paralysis, and death. The parents were frantic; the medical community was puzzled. They recognized that the deaths resulted from severe lead poisoning, but they could not figure out why the children suffered while their older siblings and parents were spared. Nor could they determine why the children who died were from the middle classes who lived in town, while the poorer children from the countryside never got sick.

It took years of investigations of water supplies and food containers before a local physician, J. Lockhart Gibson, found the answer in 1904. The children were taking in lead through the paint in their homes, paint that contained about 70 percent lead. As journalist Christopher Norwood describes in her 1980 book:

> The children of Queensland were victims of the taste and wealth of
> their parents. Because of Queensland's pounding sun and torrid days,

airy verandas, painted white from floorboard to ceiling, had become a popular addition to its town houses. These verandas, where young children would be set out to play, were the primary source of lead. Even from almost a century's distance, it is hard not to feel the poignancy of the Queensland epidemic—not to see, in the mind's eye, young mothers setting out their children in what looked to be pleasant, protected places but what were really poisonous, oversized cribs.[76]

The problem that was a middle-class phenomenon in Australia became a lower-class phenomenon in the United States, where poor children living in deteriorating slums were found in the 1960s to suffer high rates of lead poisoning, ingested from the chips and dust of leaded paint. But while overt lead poisoning quickly became a problem of the inner cities and became associated in the public mind with paint, other sources of lead were proving damaging to all children from all walks of life. The most dramatic source of lead turned out to be not paint chips but car exhaust. By the late 1960s, according to Norwood:

> so much lead was spewing from America's tailpipes and coming to rest on playgrounds or inside houses that, in some urban areas, a child need swallow only 1/24 teaspoon of local dirt or house dust to exceed the recommended daily lead intake for young children.[77]

Fortunately, since the health hazards of lead were first identified more than 85 years ago, the United States has made great strides in reducing the ambient lead levels in the air. Indeed, the reduction in lead pollution has been called one of the most significant public health success stories in recent memory. Between 1976 and 1991 the levels of lead in the blood of all Americans dropped a remarkable 78 percent. This reduction can be traced in large measure to the banning of leaded gasoline, since during this same period the amount of lead used in gasoline decreased by 99.8 percent.[78] Other environmental sources of lead—in solder in canned foods, in solder in water pipes, and in wall paint—also declined during that time. In 1991, for instance, the manufacture of lead-soldered food or soft-drink cans was outlawed in the United States.

The lead story may seem to have ended as a victory for public health activists, but it might not be over yet. While experts agree that the average blood levels among both children and adults have declined significantly in the United States, it may only be the case that lead has been driven out of the bloodstream in order to settle in a far more pernicious region: bones. Using a new technique that allows scientists to measure the element in bone, Howard Hu of the Harvard School of Public Health reported in 1994 that lead remains in the bones of adult males exposed to lead through their occupations as carpenters, demolition crew members, and other work in the construction industry. "Lead may be more toxic than we previously thought," says Hu, and "studying the ultimate toxic effects may require measuring lead that

Howard Hu with the K x-ray fluorescence instrument

accumulates in bone as well as in blood."[79] The half-life of lead in bone stores (the amount of time it takes for half of the original dose to be passed from the body) is measured in years or perhaps even decades, and it can then leach out to the rest of the body over time.

The lead problem is a graphic example of the close overlap between hazards in the environment and hazards in the social fabric of a nation, to which we now turn our attention. In the United States the first children to suffer the effects of lead poisoning were poor, living in slums where lead-based paint chips on walls and windowsills were tempting to nibble on. Outlawing lead in paints and gasoline required a deliberate social commitment; without such laws, lead poisoning could well have become one of the most intractable sources of the cycle of poverty that bedevils public health officials. If it had remained pervasive among the poor and robbed poor children of their full store of intelligence, lead could have metamorphosed from a relatively straightforward environmental hazard into a pervasive societal noose. Other examples of the health effects of the social environment—many of which are also intimately related to the physical environment—are the subject of the next chapter.

Notes

1. James H. Cassedy, "The flamboyant Colonel Waring: an anticontagionist holds the American stage in the age of Pasteur and Koch." In J. Walzer Leavitt and R. L. Numbers, eds., *Sickness & Health in America: Readings in the History of Medicine and Public Health.* Madison: University of Wisconsin Press, 1985, p. 457.

2. After Hamilton joined the faculty, it would be another 26 years before the medical school even admitted female students. The School of Public Health began admitting women in 1922, the year it was founded, but only as "special students" who were eligible to receive certificates rather than diplomas. In 1936 (according to the article "Women who made HSPH history," *Harvard Public Health Alumni Bulletin*, Fall 1982, vol. 38, p. 20), the School of Public Health finally consented to grant a degree to a woman who had finished her course work for her M.P.H. two years earlier. But even after her two-year wait, the first woman graduate of the Harvard School of Public Health was not allowed to march in the formal commencement procession from Harvard Yard.

3. Barbara Sicherman, *Alice Hamilton: A Life in Letters.* Cambridge, Mass.: Harvard University Press, 1984, p. 2.

4. Ibid., p. 238.

5. Ibid., p. 240.

6. Jean Alonzo Curran, *Founders of the Harvard School of Public Health: With Biographical Notes, 1909–1946.* New York: Josiah Macy, Jr., Foundation, 1970, p. 18.

7. In "phossy jaw" the white phosphorus used in the strike end of a match leads to loss of bone in the jaw and face, agonizing pain, physical deformity, and the emission from the victim's body of an intolerable smell.

8. "Lasker Award to Alice Hamilton," *Harvard Public Health Alumni Bulletin*, Nov. 1947, vol. 4, no. 2, pp. 31–32.

9. Theodore F. Hatch, "Expanding horizons in industrial hygiene," *Harvard Public Health Alumni Bulletin*, May 1948, vol. 5, no. 1.

10. Curran, *Founders of the Harvard School of Public Health*, p. 162.

11. Vilma R. Hunt, *Work and the Health of Women.* Boca Raton, Fla.: CRC Press, 1979, p. 208.

12. Ibid., p. 208.

13. John Little, personal interview, May 1994.

14. Ibid.

15. Joseph Brain, personal interview, Nov. 1993.

16. Leonard A. Sagan, "Policy forum: on radiation, paradigms, and hormesis," *Science*, 1989, vol. 245, p. 574. Also in the same issue of *Science*, Sheldon Wolff, "Policy forum: are radiation-induced effects hormetic?" p. 575.

17. Eliot Marshall, "Academy panel raises radiation risk estimate," *Science*, 1990, vol. 247, pp. 22–23.

18. Little, personal interview.

19. Ibid.

20. Dade Moeller, personal interview, Nov. 1994.

21. John Sedgwick, "What, me worry?" *Self*, Nov. 1993, p. 155.

22. Michael Cusack, "The lasting legacy of a steel town's ordeal," *Scholastic Update*, Nov. 1, 1985, vol. 118, p. 17.

23. Ibid., p. 17.

24. Douglas Dockery, "New thinking about ambient pollution," paper presented at "A Symposium on Respiratory Physiology and Environmental Health in Honor of James Whittenberger," Harvard School of Public Health, Feb. 11, 1994.

25. Ibid.

26. Joseph Brain, introductory comments, The James Whittenberger Symposium, Feb. 11, 1994.

27. Douglas Dockery, personal interview, May 1994.

28. Reuters, "Study ties fouled air to high urban death rates," *The New York Times*, Dec. 9, 1993, p. B15.

29. C. Arden Pope III, "Respiratory disease associated with community air pollution and a steel mill, Utah valley," *American Journal of Public Health*, 1989, vol. 79, pp. 623–628.

30. Dockery, "New thinking about ambient pollution."

31. Gregg Easterbrook, "Forget PCBs. Radon. Alar. The world's greatest environmental dangers are dung smoke and dirty water," *The New York Times Magazine*, Sept. 11, 1994, p. 62.

32. Ibid., p. 62.

33. Ibid., p. 61.

34. Ibid., p. 61.

35. George Rosen, *A History of Public Health, Expanded Edition*. Baltimore: Johns Hopkins University Press, 1993, p. 315.

36. Outbreaks of disease due to contaminated water sources have occurred in the United States in recent years, providing evidence that transmission of infectious diseases can occur in municipal water supplies. Cryptosporidium, the organism that caused illness in more than 400,000 people in Milwaukee, Wisc., in March-April 1993, was among recent wake-up calls that the safety of U.S. drinking water sources cannot be taken for granted. This has led to intensified scrutiny by U.S. public health professionals of the need for adequate water filtration to augment chlorination and changes in our approach to water supply monitoring, public health surveillance methods, and early detection of infection. See W. R. MacKenzie, et al., "A massive outbreak in Milwaukee of cryptosporidium infection transmitted through the public water supply," *New England Journal of Medicine*, 1994, vol. 331, pp. 161–167. Emerging water-borne pathogens are discussed in Timothy Ford and Rita R. Colwell, "A global decline in microbiological safety of water: a call for action," American Academy of Microbiology, Washington, D.C., in press.

37. Douglas Hand, "The long struggle to be heard," *American Health*, May 1988, p. 51.

38. Ibid.

39. Marvin Zelen, personal interview, Nov. 1993.

40. Ibid.

41. Paula DiPerna, "Leukemia strikes a small town," *The New York Times Magazine*, Dec. 2, 1984, p. 102.

42. Jonathan King, *Troubled Water*. Emmaus, Pa.: Rodale Press, 1985, p. 133.

43. Phil Brown, "Popular epidemiology and toxic waste contamination: lay and professional ways of knowing," *Journal of Health and Social Behavior*, 1992, vol. 33, p. 272.

44. Brian MacMahon, "Comment," *Journal of the American Statistical Association*, 1986, vol. 81, pp. 597–599, one of four comments following the article under discussion, Stephen W. Lagakos, B. J. Wessen, and Marvin Zelen, "An analysis of contaminated well water and health effects in Woburn, Massachusetts," *Journal of the American Statistical Association*, 1986, vol. 81, pp. 583–596.

45. Stephen Lagakos, personal interview, Oct. 1995.

46. The word from Hollywood in late 1995 was that Robert Redford was interested in turning *A Civil Action* by Jonathan Harr (New York: Random House, 1995) into a full-length feature movie.

47. National Institute of Dental Research, National Institutes of Health, Bethesda, Md.

48. Elizabeth Fee and Barbara Rosenkrantz, "Professional education for public health in the United States." In E. Fee and R. M. Acheson, eds., *A History of Education in Public Health: Health That Mocks the Doctors' Rules*. Oxford and New York: Oxford University Press, 1991, p. 238.

49. Fitzhugh Mullan, *Plagues and Politics: The Story of the United States Public Health Service*. New York: Basic Books, 1989, p. 112.

50. Robert L. Crain, Elihu Katz, and Donald B. Rosenthal, *The Politics of Community Conflict: The Fluoridation Decision*. Indianapolis: Bobbs-Merrill Co., 1969, p. 16.

51. Donald R. McNeil, *The Fight for Fluoridation*. New York: Oxford University Press, 1957, p. 67.

52. Ibid., p. 19.

53. Ibid., pp. 22–23.

54. Edward O'Rourke, "Fluoridation—the case for," *Harvard Public Health Alumni Bulletin*, June 1953, vol. 10, no. 1, p. 10.

55. Rachel Carson, *Silent Spring*. New York: Fawcett Crest Edition, 1967, p. 168.

56. John Graham, personal interview, July 1994.

57. "Sweeteners: are any of them safe?" *Consumer Reports*, Nov. 1985, p. 691.

58. Graham, personal interview.

59. Robert Meyers, *DES: The Bitter Pill*. New York: Seaview/Putnam, 1983, p. 15.

60. Ibid., p. 16.

61. Diana Dutton, *Worse Than the Disease: The Pitfalls of Medical Progress*, Cambridge: Cambridge University Press, 1988, p. 31, quoting an advertisement from the Grant Chemical Co. that appeared in the June 1957 issue of the *American Journal of Obstetrics and Gynecology*.

62. Meyers, *DES: The Bitter Pill*, p. 93.

63. Ibid., p. 94.

64. Ibid., p. 98.

65. Ibid., p. 96.

66. Arthur L. Herbst, Howard Ulfelder, and David C. Poskanzer, "Adenocarcinoma of the vagina," *The New England Journal of Medicine*, 1971, vol. 284, pp. 878–881. The article was accompanied by an editorial by Alexander D. Langmuir, "New environmental factor in congenital disease," pp. 912–913, who called it "an original publication of great scientific importance and serious social implications." Indeed, as Dutton describes in *Worse Than the Disease* (p. 71), the editor of the *NEJM* considered the report so urgent that he broke his own prepublication news blackout to send a draft of the article to the U.S. Food and Drug Administration (FDA) as soon as it was accepted. The FDA responded by requesting the raw data from Herbst, who promptly complied. But no one heard again from the FDA on this matter for another eight months—during which time physicians, unaware of the risks, wrote prescriptions that exposed another 60,000 or so fetuses to the harmful effects of DES.

67. Peter Greenwald, Joseph J. Barlow, Philip C. Nasca, et al., "Vaginal cancer after maternal treatment with synthetic estrogens," *The New England Journal of Medicine*, 1971, vol. 285, pp. 390–392.

68. Meyers, *DES: The Bitter Pill*, p. 145.

69. A summary of the health of DES sons can be found in William B. Gill, Gebhard F. Schumacher and Marian M. Hubby et al., "Male genital tract changes in humans following intrauterine exposure to diethylstilbestrol," in A. L. Herbst and H. A. Bern, eds., *Developmental Effects of Diethylstilbestrol (DES) in Pregnancy* (New York: Thieme-Stratton, 1981, pp. 103–119).

70. Dutton, *Worse Than the Disease*, p. 61.

71. The DES ban in chicken feed occurred, according to Dutton (p. 61), after a study showed that DES residues in marketed poultry were 342,000 times the levels found to be carcinogenic in mice.

72. Dutton, *Worse Than the Disease*, p. 62.

73. Ibid., p. 77.

74. Ibid., p. 78.

75. In *Worse Than the Disease*, p. 80, Dutton quotes from Keith Schneider, "Congress looks to the American table amid questions of food safety," *The New York Times*, June 22, 1987, p. 10, who summarized several studies that provided "more convincing evidence than ever before that modern food production, and the inability of Federal agencies to adequately inspect food and detect toxic substances, may be jeopardizing Americans' health."

Confronting the Physical Environment

76. Christopher Norwood, *At Highest Risk: Protecting Children from Environmental Injury.* New York: Penguin Books, 1980, p. 104.

77. Ibid., p. 106.

78. James L. Pirkle, Debra J. Brody, and Elaine W. Gunter et al., "The decline in blood lead levels in the United States," *Journal of the American Medical Association,* 1994, vol. 272, pp. 284–291.

79. "HSPH study examines bone lead toxicity," *Around the School: News & Notices of the Harvard School of Public Health,* Nov. 18, 1994, p. 2. See Howard Hu et al., "The relationship between bone lead and hemoglobin," *Journal of the American Medical Association,* 1994, vol. 272, pp. 1512–1517.

The
People's
Health

4

Confronting the Social Environment

Ⅰf you were a resident of Alameda County, California, in 1965, you might have received a visit from a stranger that was to help shape the course of public health survey research. That year investigators from the University of California at Berkeley knocked on 4,735 doors in Berkeley, Oakland, and the surrounding suburbs to distribute questionnaires for what has come to be known as the Alameda County Study.

For the first time, researchers set about systematically to correlate familial, cultural, economic, social, and environmental factors with physical health. They ended up receiving completed questionnaires from nearly 7,000 adults over the age of 20. After evaluating the responses, and the responses to a follow-up survey conducted in 1974, the Berkeley investigators discovered that the social environment was just as important as any other factor in determining who got sick and who did not.

Investigators correlated three social variables—race, income, and residence in a poverty or nonpoverty area—with their subjects' health status in 1965 and again in 1974. As expected, the people who enjoyed the best health were those with higher incomes living in nonpoverty areas. But, surprisingly, the immediate social community had such a profound effect on health that even at the same level of income, people who lived in poverty areas had more health problems than those who didn't.[1]

To investigate the impact of social integration or isolation on general health and longevity, investigators asked three questions:

1. How many friends do you have?
2. How many relatives do you have that you feel close to?
3. How often do you see these people each month?

According to Lisa Berkman of the Harvard School of Public Health, who began analyzing these questions as a graduate student working with Leonard Syme at Berkeley in the 1970s, people with few social contacts were up to three times more likely to die than those with stronger social networks. As she explains:

> Considered individually, none of these questions were important predictors of mortality. However, when combined, they are associated with significant increases in risk. In every age and sex category, people who report having few friends and relatives and/or who see them infrequently have higher mortality rates than those people who have many friends and relatives and see them frequently.[2]

Berkman developed a Social Network Indicator to summarize the effects on mortality of increasing social isolation. "It considers not only the number of social ties but also their relative importance," she says. "Thus, intimate contacts are weighted more heavily than church affiliations and group memberships." The protective effect of intimate social contacts holds true even when other factors, such as physical health, socioeconomic status, and use of preventive health services, are taken into account.[3]

Though the Alameda County study was the largest and longest-running study of its kind, investigations into social conditions and health had begun more than a century before the Berkeley researchers first got together. In 1846 the physician Rudolph Virchow studied the typhus epidemic in Germany. He found that crowded conditions among the poor led to higher rates of typhus, while wealthier Germans who lived in larger houses with fewer inhabitants tended to be spared. In 1850 Lemuel Shattuck, a Boston bookseller, educator, and proponent of the collection of vital statistics, wrote a groundbreaking paper that stressed the need to improve the health status of all residents of Massachusetts, particularly those living in poverty.[4] And in the early 1920s a survey of more than 8,500 residents of Hagerstown, Maryland, highlighted the complex relationship between morbidity (illness) and mortality (death). During the 28 months in which the residents were fol-

Lisa Berkman

lowed, the illnesses that made them sick were quite different from the ones that tended to kill them.[5]

Yet it was not until the 1950s, when the Alameda County Study was about to get under way, that the field of public health began to appreciate the impact of the social environment. The work of Benjamin Paul, who later joined the faculty of the Harvard School of Public Health, suggested that the relationship between health and society deserved more attention.[6] Arguably the "granddaddy" of the Alameda County Study and others to come was the Stirling County Study, a description of psychiatric disorders in relation to sociocultural factors in a population in Nova Scotia. Planned and implemented in the mid-1950s by Alexander Leighton, who also later became a member of the faculty at the Harvard School of Public Health, Stirling County was the first survey of a population to propose a theory of health in relation to culture.[7]

In his 1957 presidential address to the American Public Health Association, John W. Knutson emphasized the need to train an emerging generation of public health leaders in a broader view of health, one that encompassed the social conditions under which people lived.

> Our graduate students need a better understanding of the social and political forces swirling about them as they work professionally. In place of rapidly outdated factual information, I would have them gain from graduate education in public health a fundamental knowledge of four broad fields: cultural anthropology, human ecology, epidemiology, and biostatistics.[8]

Epidemiology and biostatistics had long formed the intellectual core of the field of public health. It was Knutson's inclusion of the two other fields—cultural anthropology and human ecology—that marked the real sea change in the general conception of public health and the scope of disciplines considered able to contribute to its advancement.

The new discoveries and methods applied to social issues brought with them a need to expand and broaden the generally accepted notion of public health's mission. This expansion had to occur in several directions: not only deep—within each scholarly discipline—but also wide—across entire fiefdoms of academic departments. "The new community health problems were not encompassed by traditional biomedical concepts," writes historian John Duffy, describing the new view of public health in the years just after World War II.

> Alcoholism, drug addiction, smoking, radiation, environmental hazards, and the problems of aging scarcely fitted the normal categories. . . . The concept of mental illness, an amorphous term at best, made the whole area of human emotions a legitimate sphere for public health. The virtual elimination of the major communicable diseases in the twentieth century had been relatively simple and cheap; solutions to the new public health problems, however, involved social reform and large public expenditure.[9]

By the 1980s, schools of public health had developed specific programs to help scholars address the complex interaction between society and health. Through these varied programs investigators sought to answer one common question: Why do so many disorders that ultimately express themselves as health problems have their origins in social disruptions and social forces?

"Social factors are of overwhelming importance in determining health," says Sol Levine, who, with Diana Chapman Walsh, helped organize the Society and Health Program at the Harvard School of Public Health. By social factors, Levine says,

> I mean not only poverty but also social class, family, community, gender, ethnicity, racism, political economy, and culture. We have to learn how these interact with health, how, for example, culture, political economy, and racism may affect the community and family environment, which, in turn, may influence people's health. We have to look not only at individual characteristics but to the features of the society as well.[10]

For example, says Levine, there is reason to believe that in the developed countries the degree of social and economic inequality—the distance between the haves and the have-nots—is strongly related to life expectancy. Developed societies that have more equitable distribution of income seem to have the highest life expectancy.

This concept has been quantified as the "Robin Hood Index" and was employed by three scientists from the Harvard School of Public Health in a recent study. Bruce Kennedy, Ichiro Kawachi, and Deborah Prothrow-Stith used the index to represent the share of total income that has to be taken from people above the average income and transferred to those below the average in order to achieve equality in the distribution of income. They calculated such a number for each of the 50 states and, using mortality and morbidity statistics from U.S. data banks, found that the Robin Hood Index explains more than half the variation in the homicide rate among different states. It was also significantly associated with variations in deaths from heart disease and cancer.[11]

This understanding of the re-

Ichiro Kawachi

lationship between society and health forms the backbone of this chapter. While the stories contained here might at first seem unrelated, what they share is a particular societal irregularity that manifests itself as a problem in public health. That irregularity may be rooted in economics and science policy, as in the case of mass vaccination programs that have worked and failed to work over the past 50 years; in governments, as in the case of the medical consequences of a nuclear war; or in the intimate institutions of family, school, home, and neighborhood, which have conspired to create an epidemic of violence in many American cities in the 1990s. Each of these cases will be considered in turn.

Overriding all these social strains is one salient factor: poverty. It is to this basic disruption of the social fabric that we turn our attention first, to provide a backdrop for understanding the other social ills that ultimately play themselves out in the arena of public health.

Poverty and Disease

Quite apart from the dreadful life circumstances that accompany poverty—inadequate housing, malnutrition, inattention to preventive care, stress, exposure to drugs and violence—the very fact of being poor is itself an independent risk factor for getting sick. It is simply unhealthy to be poor, points out William Foege of the Carter Center in Atlanta:

Confronting the Social Environment

> In almost any study that you can find in any culture, mortality rates are higher for almost every category of disease in the poor as compared to the rich. And it's not just that we have to take the very poor and the very rich. There's a gradation in between. There's something about poverty that makes it a health problem.[12]

No one has yet figured out the precise connection here. But even when you correct for any of the confounding factors that might account for the difference—cigarette smoking, alcohol use, drug use, access to health care services—income still comes out as an independent predictor of health status. Foege thinks it has something to do with a sense of fatalism among the poor.

> It's very easy for the rich and the educated to have control of every new thing that comes out. Something's in the newspaper, and 12 hours later people are already starting to apply it. The poor don't have that option.[13]

Observations of the link between poverty and disease go back many years, but a century or more ago the connection was taken to reflect a moral weakness on the part of the poor and sick. In 1891 John Shaw Billings, surgeon general of the United States, made an important epidemiological finding: the death rates of various hospital wards varied in inverse proportion to the income of the patients in those wards. In other words, he found that the lower the income of its patients, the higher a ward's death rate. Billings drew from this a conclusion that today seems shocking indeed: he said it reflected the moral failings of "a distinct class of people who are structurally and almost necessarily idle, ignorant, intemperate and more or less vicious, who are failures or the descendants of failures."[14] Thankfully, we have come a long way from this dangerous reasoning. A modern analyst would reach quite a different conclusion from Billings: that the poor have higher death rates because they have less access to good medical care, good nutrition, community resources, and a healthful sense of control over their own destinies.

Being poor affects health in many ways. One of the most obvious is that it keeps people from getting the medical treatment they need. Low-income people often cannot afford cab fare, cannot afford medicine, cannot afford screening tests, and do not have health insurance anyway. It takes more emotional and intellectual resources than many impoverished families can summon to figure out how to get through the maze of clinics, personnel, and paperwork that are part of an encounter with the fragmented health care system available to most low-income people. Access to care was even more restricted for the poor prior to the early 1960s and the Great Society programs, when Medicaid removed at least one barrier—an inability to pay. But all too often, problems with access go beyond finances.

This was demonstrated in 1966 by Alonzo Yerby, the New York City hospitals commissioner who was soon to return to his alma mater, the Harvard School of Public Health, as a faculty member. Yerby spent some time with a family on public assistance that comprised four children, each of whom suffered from a chronic disease. In this family there was no possibility of cutting back on doctor visits; the children were too seriously ill for that. In a typical week, Yerby found, the mother had to cart her children to 11 appointments at eight different outpatient clinics at five different hospitals—a logistical nightmare for even the most resilient family. The local health department, informed of the discontinuity of care the family faced, made no effort to consolidate health care services. City health officials simply sent an irate letter to the welfare office demanding that the family receive an increase in its transportation allowance.[15] As Yerby wrote at the time:

> Health care of the disadvantaged is many things. It is frequently inadequate, sometimes quite poor, provided with little dignity or compassion, and rarely related to the total needs of the individual or family. . . . The pervasive stigma of charity permeates our arrangements for health care for the disadvantaged, and whether the program is based upon the private practice of medicine or upon public or non-profit clinics and hospitals, it tends to be piecemeal, poorly supervised, and uncoordinated.[16]

Poverty also has some not-so-obvious effects on health. Ghetto housing places people at higher risk for environmentally induced conditions, such as asthma (associated with exposure to cockroach droppings in the home), tuberculosis (associated with crowded living quarters), and lead poisoning (from exposure to peeling lead-based paint and from tap water that passes through old service pipes, which bring water from the city supply into the home and are almost always made of lead). In the poorest neighborhoods, people are more likely to encounter drug and alcohol abuse, domestic violence, and street crime. Five times as many black women as white women, for example, are victims of homicide.[17]

Many poor families must make decisions every day about how to stretch their impossibly stretched budgets. Often, good food is the first thing to be sacrificed—helping assure that the next generation, too, will grow up hampered. The paradox of poverty and food was first revealed in 1936 in a classic public health study conducted in the British town of Stockton-on-Tees. The British government had made an attempt to lift the poor out of squalor by removing slum dwellers from their dilapidated housing and placing them in new, more hygienic, government-issue flats.

But surprisingly, the health status of the relocated population promptly declined. Within a few years, public health officials were able to chart an increasing death rate among people in the new housing, as compared to their former neighbors still living in the ghetto. "They could actually be worse off in new housing," wrote historian George Rosen, "because more of the inadequate family income had to be spent for rent and less was left over for food."[18]

In the years since, study after study has documented the shameful health care gap between society's haves and its have-nots. As Duffy writes:

> Public health is determined to a large extent by the public's standard of living. A well-fed and properly housed population is far more resistant to all of the ailments besetting humanity than one that is impoverished. The reduction in infant mortality and the increase in life expectancy in England during the eighteenth century was helped by the work of reformers, but it rested on the country's increasing general wealth. In the United States today the higher maternal and infant mortality rates and lower life expectancy among most minority groups reflect their below-average income levels. And poverty, regardless of race, invariably breeds sickness.[19]

Many biostatisticians use the terms "poor" and "minority" almost interchangeably—or, as they put it more technically, use race as a "surrogate" for income. They feel justified in doing so because poverty rates are significantly higher for minorities. Among African-Americans, for instance, one-third are living in poverty, compared to just 11 percent of whites. Among African-Americans under age 18, the gap is even wider: 47 percent of black children live in poverty, compared to 16 percent of white children.[20]

"Yes, race is a rough proxy for socioeconomic status," says Camara Jones of the Harvard School of Public Health. "But it is not a wonderful proxy, because although a disproportionate number of black people are poor, the majority of poor people in this country are white, and not all black people are poor."[21]   And when you assume that race explains as many health differences as does poverty, she says, you miss one critical social factor in determining a public health problem: racism.

When epidemiologists detect race-associated differences in a particular health outcome—high blood pressure, for example—they usually factor in socioeconomic status to account for differences that might have been explained by income or education, says Jones. Then when they still see differences from one race to another, they assume that whatever is left is caused by genetics. But Jones says those differences are a graphic measure not of genetics but of racism.

> We live in a race-conscious society, where the way somebody looks determines their life experiences. What we measure when we measure race—or even have people self-identify by race—is basically a social construct that's reflecting that fact. I see racism operating on three levels, and each one has an effect on health. There's institutionalized racism, which is the historical legacy of differential access to education, jobs, and that kind of thing; personally mediated racism, when somebody makes judgments about another person's ability, motivation, or intent on the basis of that person's so-called race; and internalized racism, where stigmatized people internalize the negative messages they keep hearing about themselves.[22]

As a first step in quantifying how racism directly affects a person's health, Jones inserted a question into the 1995 version of the Nurses' Health Study (described in Chapter 5) to measure internalized racism. The question is: "How often do you think about your race: never, once a year, once a month, once a week, once a day, once an hour, once a minute, or constantly?" She says her experience is that whites (who make up about 97 percent of the Nurses' Health Study population) tend to respond in the range of "never" to "once a week," while blacks tend to respond somewhere between "once a week" and "constantly." The next step is to see whether thinking about race more leads to adverse health outcomes.

One of the most sensitive measures of the relative health of a society is infant mortality, in large part because women and developing fetuses are highly vulnerable to environmental conditions, such as housing, sanitation, and food and water, which are compromised among the poor.[23] In the United States, which has a shamefully high rate of infant mortality compared to other industrialized countries, the income gap in infant mortality is enormous.[24] In a census tract in the racially mixed Mission Hill neighborhood of Boston, located across the street from the Harvard School of Public Health, the average death rate in the first

year of life during the late 1970s was 47 per 1,000 births—three times the national average.[25]

The conventional wisdom says that this social and economic disparity in infant mortality in the United States is caused by a lack of good prenatal care for poor women. But that doesn't really explain it, since the conditions that poor women's babies die of—primarily prematurity and very low birthweight—are not generally preventable through obstetrical visits alone. As Marie McCormick of the Harvard School of Public Health says, the reason for high neonatal death rates among the poor is the unraveling of all the safety net programs since the early 1980s. It is not simply lack of prenatal care.

> You have to understand what prenatal care was designed to do. Prenatal care originally was designed to manage the problem of toxemia of pregnancy [dangerously high blood pressure in the last few months]. That is essentially why the schedule is the way it is, with relatively few visits early on and most of the visits concentrated in the third trimester. The rate of second-trimester deliveries, true preterm infants, has not changed in 30 years. And it's not clear why it would as a result of improved prenatal care, where the window of opportunity for intervention between the start of prenatal care and very preterm delivery may only be a matter of weeks. [26]

Maternal mortality is another risk borne disproportionately by the poor. In the years between 1940 and 1985 the national rate of pregnancy-associated mortality dropped significantly, from 376 per 100,000 live births to just 7.8. But most of the advances were limited to women of the middle and upper classes. "National rates are low only for whites,

107

with non-white rates being three times higher," wrote researchers with the New Jersey State Department of Health in 1992. "Furthermore, upon review, one-third to one-half of pregnancy-related deaths are considered preventable. Finally, deaths from AIDS have begun to increase the pregnancy mortality ratio in some areas."[27] So even something we thought had been relegated to the dark memories of the past—women dying during or just after childbirth, leaving behind motherless babies— is still, for too many poor Americans, a continuing fact of life.

Both racism and poverty often lead to long-term health deficiencies that make it all but impossible for minorities, especially low-income minorities, to catch up. "Much of the medical damage resulting from malnutrition, unmonitored pregnancies, early childhood infections, and other blights often associated with poverty can never be repaired by even the most sophisticated forms of medicine,"[28] Howard Hiatt, then dean of the Harvard School of Public Health, wrote in 1982, at a time when the administration of President Reagan was imposing deep budget cuts on Medicaid and other health and welfare programs for the poor. Those conditions that medicine cannot repair, he wrote, public health must try to prevent.

Prevention of the ills of poverty is at the heart of the Project on Human Development in Chicago Neighborhoods, an interdisciplinary study directed by Felton Earls of the Harvard School of Public Health. The project, which tracks the development of 9,000 children and adolescents in 80 Chicago neighborhoods, is designed to detect the individual and community characteristics that differentiate success from failure for those young people. As Earls puts it:

> I could take a daring, impulsive 9 year old in one neighborhood and compare him with a kid who is exactly as daring and impulsive but who lives in another neighborhood. The difference between these neighborhoods is that one is poor, one is affluent; one has easy access to drugs and crime because it's disorganized, and the other has softball teams, Boy Scout troops, after-school programs, music lessons, where adults are supervising kids and encouraging skills development throughout their adolescence. In the disorganized community, gangs seem to be the substitute for the Boy Scouts and the after-school programs. And so what happens to this 9-year-old daring kid in the gang-run neighborhood is altogether different than what happens to the kid in the more organized one. Therefore, it's not the kids alone that determine their future. It's the influences and opportunities that surround them.[29]

Although affluent neighborhoods tend to be better organized than poorer neighborhoods, the relationship is not always direct. "Once you get above the poorest of the poor neighborhoods," says Earls, "where you have upper-working-class to lower-middle-class incomes in the $20,000 to $40,000 range, that's when you start to see a lot of variation" in the degree of social organization and the ultimate outcome of the children. There are strong poor neighborhoods just as there are weak affluent neighborhoods, and Earls thinks it may be the strength of the social organization—rather than the average income of the residents—that determines whether kids stay in school and go on to college and careers or drop out and become drug addicts and criminals.

Earls points to a reason for the epidemic of violence plaguing America over the past decade: "Our hypothesis is that Chicago's neighborhoods are changing systematically across all classes and races," he says. "And the major way in which they're changing, we believe, is that adults are becoming less committed to children."[30]

As adults give up on children, children respond by turning to crime, drugs, alcohol, guns, and, eventually, violence. Only in the past decade have public health experts taken on the epidemic of violence-related injury and death as a problem that is within their power to prevent. Prospective descriptive studies of high-risk youngsters, of which Earls's is one of the largest, are one way in which this new mandate is being carried out. We will turn now to other efforts that tend to share a common theme: an attempt to apply traditional public health tools to America's newest, most pernicious epidemic—violence.

Violence as a Public Health Problem

The young medical student worked carefully on her patient, a stab wound victim in the Boston emergency room where she was doing her surgical rotation. It was 3 o'clock in the morning, and the patient's deep cut was just an inch above his eye. The medical student apologized, as she did to all her other patients on that long winter night in 1978, for her slowness at stitching. "I'm still in training," she explained.

"Don't worry," said the young man. "By the end of tonight you'll be a perfect doctor. You'll get all the practice you need." He bragged that as soon as she fixed him up, he was going back on the streets to look for his attacker—and he intended to hurt him so badly that the guy who started it all would end up in the emergency room, too. He advised her not to bother going to sleep.

It was an epiphany for the young medical student, who realized that the violence of Boston's city streets was every bit as much of a health hazard as the cigarettes people smoked or the high-fat diets they ate. But here she was, stitching up a victim of violence only to send him back out to create a new victim, who would also need treatment—a kind of revolving-door medicine.

"We never sent a heart attack victim home to have another heart attack; we taught them about diet, stress, exercise," says the former medical student, Deborah Prothrow-Stith, now an assistant dean at the Harvard School of Public Health and a national leader in the effort to confront violence as a public health issue. "We never sent a suicide-attempt home to try again. We always taught prevention. Except for

Deborah Prothrow-Stith

victims of violence."[31] Since that night, Prothrow-Stith has lent her voice to the growing movement to force public health experts to confront violence as they would any other epidemic: by studying its causes, prevention, and cure.

One of the most prominent people to define violence as an epidemic and to suggest a public health approach to its prevention was C. Everett Koop, the renowned surgeon general from 1981 to 1989. In 1984 Koop announced:

> Violence is as much a public health issue for me and my successors in this country as smallpox, tuberculosis, and syphilis were for my predecessors in the last two centuries.[32]

Despite Koop's popularity and his ability to use his position as a bully pulpit to rally the nation to confront AIDS and cigarette smoking, his proclamation regarding violence received scant attention. Nor was much attention paid in 1983 when the Centers for Disease Control (CDC) declared firearm-related violence a public health hazard and set about studying it with the same epidemiological tools used in earlier epidemics to study toxins or pathogens.[33]

Not only was the public health approach to violence prevention ignored, in some instances it was actively resisted, sometimes from the public health community itself. Prothrow-Stith was stunned when her colleagues told her in the mid-1980s that violence was beyond their purview—an attitude she found especially perplexing in light of other changes that were taking place simultaneously in the field of public health.

110

Doctors were playing an increasingly active role in many areas related to health that had little to do with disease in the old-fashioned sense. My colleagues were fighting in favor of laws requiring children to ride in special car safety seats. They were campaigning in favor of mandatory seat belts. They were involved in efforts to curb smoking and impede drunk driving. They were screening patients in hospitals for battery and child abuse. Physicians were broadening their base, defining the pursuit of good health in new and imaginative ways. To my mind, violence prevention fit right in with all these other forms of "health promotion."[34]

But as the decade progressed, violence became so pervasive that anyone could see the urgency in trying to develop ways to prevent it. Death rates from violence, especially among the young, were soaring. Among Americans aged 15 to 34 the death rate from homicide increased by 50 percent between 1985 and 1991, from 13.4 deaths per 100,000 people to 20.1. During that time, the rate of increase was three times higher for the subset of young people who were male teenagers. In this group (15- to 19-year-old males) the homicide death rate between 1985 and 1991 skyrocketed by 154 percent.[35]

Gradually, evidence accumulated about the factors that helped contribute to this appalling trend. Chief among them was the spread of firearms. As the 1990s dawned, scientists felt they had demonstrated a clear connection between the availability of firearms and an increase in the homicide rate—a connection that was, as one researcher put it, "every bit as strong as the studies that linked cigarettes to lung cancer."[36]

One piece of evidence came from a study reported in the late 1980s by Arthur Kellerman of the University of Tennessee. Kellerman compared crime statistics for 1980 through 1986 from two cities: Seattle and Vancouver. He chose the two cities because they were comparable on many measures—geography, climate, socioeconomic characteristics, histories, culture, even local television shows—but varied significantly in one salient respect. Firearms were easy to obtain in Seattle but, because of strict Canadian gun control laws, were virtually nonexistent in Vancouver. The two cities turned out to have similar amounts of criminal activity as measured by rates of burglary, robbery, and assault. But the homicide rate was 60 percent higher in Seattle, and the rate of homicide by firearms was a staggering 500 percent higher.

In an editorial in *The New England Journal of Medicine* in which Kellerman's findings were published, two CDC investigators wrote:

> The time has come for us to address this problem in the manner in which we have addressed and dealt successfully with other threats to public health.[37]

This did not necessarily mean completely understanding all the factors that lead to violence any more than an assault on tuberculosis in the late 1800s required a complete understanding of the pathogen that caused it. Just as tuberculosis control began long before the tubercle bacillus was identified—indeed, before it was even hypothesized—so did violence control begin even in the absence of any good data about its predisposing factors, underlying causes, or coexisting risks. As Earls points out:

> The prevalence [of tuberculosis] steadily declined over an entire century before specific therapeutic and preventive agents were introduced. The public health lesson worth repeating is that we have often devised methods of controlling a disease with poor, and sometimes even misleading, ideas about its true causation.[38]

Once the public health approach to violence prevention was adopted, it became possible to make useful analogies to controlling epidemics of tuberculosis and other diseases. One of Prothrow-Stith's favorites is the comparison between the prevention of violence and the prevention of lung cancer. Each effort involves three levels of intervention, she says: primary (health messages and health promotion campaigns), secondary (intervention), and tertiary (treatment). In terms of lung cancer, primary prevention efforts through the 1960s and 1970s effectively turned smoking from a sophisticated habit to what Prothrow-Stith calls a "repulsive behavior." Similarly, antiviolence messages can create a new definition of acceptable standards of behavior. While an earlier idea of "cool" might have been personified by the guy who stands his ground and fights, primary prevention efforts can make it seem more cool to be the one who takes a deep breath and walks away from an argument before it escalates.

Secondary prevention for lung cancer involves smoking cessation efforts. In violence prevention, secondary prevention means teaching schoolchildren how to deal with their anger and enhancing their sense of self-worth through mentor programs and after-school activities. Finally, tertiary prevention for lung cancer means surgery or drug therapy; for violence prevention it means prison. Tertiary prevention is a last-ditch effort, says Prothrow-Stith, to be resorted to only when primary and secondary prevention efforts fail. "In violence prevention as in lung cancer prevention," she writes, "the early intervention strategies are smarter, cause less pain and are cheaper."[39]

By 1993, when David Satcher was named director of the CDC, it was almost a given that violence was very much within the province of public health. Articles about Satcher's philosophy carried such headlines as "CDC's New Chief Worries as Much About Bullets as About Bacteria."[40] Satcher made it clear that preventing injury, both intentional and unintentional, was one of his top priorities:

> If you look at the major cause of death [in young Americans] today it's not smallpox or polio or even infectious diseases. Violence is the leading cause of lost life in this country today. If it's not a public health problem, why are all those people dying from it?[41]

Violence takes its toll not only in lives lost or medical costs incurred but also in psychological damage. In fact, the threat of violence, which often permeates an entire population's way of thinking, is often greater than the actual loss of life to violence in a community. In 1993 Prothrow-Stith documented the psychic costs of living in a violence-plagued neighborhood by interviewing more than 2,000 students in grades 6 through 12. Among the statements she asked youngsters to agree or disagree with was this: "My chances of living to a ripe old age are likely to be cut short because of the threat of my being wiped out from guns." One out of three youngsters said they agreed.[42]

The very existence of violence in a community can pervade people's lives in ways that haunt their every movement. Consider the words of Mukiya Adams, 17, whose essay appears in Prothrow-Stith's book, *Deadly Consequences*:

> Each day when I come home from school, I stay in the house until the next morning, when it's time to go to school again. When I leave school in the afternoon, I try to find a safer way to walk home. But there's no safe way to come. No matter which way you come, you always have that fear inside.[43]

Children like Mukiya, a high school student in Boston, are suffering from a psychological syndrome that Prothrow-Stith compares to the posttraumatic stress syndrome of war veterans. "The one distinct difference is [that] the veterans have come home," she says. "These children are still on a daily basis dealing with the fears, the anxiety and the pain."[44]

When teenage homicide is viewed from an epidemiological perspective, public health professionals have been able to show the ways in which young people who kill are, quite often, almost interchangeable with their victims. Both are likely to be male; poor; of the same race; depressed (although not necessarily psychotic); to have been exposed to violence in the past, either directly or as witnesses; and to use or abuse drugs or alcohol.[45] Once epidemiologists identify the individuals most likely to be touched by teenage violence—whether as aggressor or victim—they can join with their public health colleagues in other specialties to mount preventive campaigns in schools, homes, and communities. This involves efforts such as Prothrow-Stith's violence prevention curriculum, which she developed and has introduced into elementary and middle school classrooms in over 500 school systems nationwide. Or it means making handguns childproof—perhaps by personalizing them so that only their registered owners can shoot them—or developing bullets that maim without killing.

But change, whatever shape it takes, will occur slowly.[46] "What public health lacks is a sense of urgency," writes Prothrow-Stith. "Scientists trained to analyze the data carefully and dispassionately are not the best advocates for changing the world."[47] Nor is human behavior especially malleable. It took 25 years of constant hammering on the part of public

113

health professionals at all levels of government to reduce the smoking rate in the United States by as much as 30 percent.

The ultimate violence, of course, and the one most difficult to prevent, is violence of global rather than neighborhood dimensions—the weapons of mass destruction that have been stockpiled by the world's superpowers for the past 50 years. Preventing nuclear annihilation requires an entirely different set of skills from prevention of hand-to-hand violence, but it too is a challenge that a few brave souls from the public health community have defined as part of their mission.

The Cold War and the "Last Epidemic"

In retrospect it seems a short hop from concern about premature death from preventable medical conditions to concern about the most sudden, most devastating death of all—death from nuclear war. That at least is how Bernard Lown sees it. Lown has had two consuming passions during his lifetime—studying sudden death (a cardiac condition in which the first and only symptom of heart trouble is when the patient drops dead) and stopping the nuclear arms race—which he sees as fully complementary, with the second interest the inevitable outgrowth of the first. What death could be more sudden than death from an atomic bomb?

Lown's involvement in the antinuclear movement began in 1961 when he was trying to interest his colleagues at the Harvard School of Public Health in sudden death. "It was a huge problem but largely ignored as a public health issue," he recalls. His good friend Roy Menninger, who was a resident in psychiatry at the Massachusetts General Hospital, invited Lown to a lecture by Philip Noel-Baker, a British diplomat who had received the Nobel Peace Prize for his role in the founding of the League of Nations.

"What is he going to talk about?" Lown remembers asking.

"The nuclear threat," said Menninger.

"Forget it," Lown replied; "I have enough to do." But Menninger was persistent, and Lown eventually went to the lecture, held in an elegant home in Cambridge. There Lown sat among 50 or so strangers and was transfixed.

> Noel-Baker was like an ancient Hebrew prophet, like some Jeremiah intoning about the fact that the world would not be around for long because the forces at work then were beyond human control. The nuclear missiles were increasingly automated, increasingly responsive to one another, their destructiveness was awesome, unprecedented, and would put an end to civilized society as we knew it. The vision he conjured was so powerful that I went back home and I couldn't sleep.[48]

Within weeks Lown had called together a group of about a dozen doctors, most of them affiliated with the Harvard School of Public

Health or the Harvard Medical School, to meet in his living room in Newton, a suburb of Boston. "I said to them that doctors have a responsibility to the health of their community," he says, "and the greatest threat to the health of their community was not anything else but the nuclear threat. And everybody bought that. It wasn't hard to sell."

Soon the informal study group, which adopted the name Physicians for Social Responsibility, was meeting every two weeks in Lown's home, focusing their developing expertise on one particular subject: the medical consequences of nuclear war.[49] The group tried to calculate the local effects of a 10-megaton bomb directly hitting the city of Boston.[50] Two million of Boston's 3 million citizens would be killed instantly, they calculated, with another 500,000 mortally injured. Among the dead would be 5,000 of the city's 6,000 physicians. "Then," recalls Lown, "we said, 'What would those 1,000 remaining doctors do?'"

> No electricity, no gas, no transport, unburied bodies, epidemics raging, fires, radiation. The whole triage system becomes meaningless. Let's say you have a person walk in and he has a scratch. Should you attend to him if he has already had a lethal body dose of radiation, rather than tending to the guy who's critically injured but has no radiation exposure? How would you know that one is going to die whatever you do, while the other would have lived but you wouldn't pay attention to him because he looked so badly wounded? What we concluded is that medicine would become completely irrelevant. Doctors would have nothing to offer except perhaps to practice euthanasia.[51]

Such a conclusion was revolutionary. Despite earlier government documents describing postbomb triage, never before had the aftereffects of a nuclear detonation been described so graphically. Never before had the folly of America's so-called civil defense program—with fallout shelters that would turn into vacuums and suffocate their occupants and a "duck-and-cover" mentality that would be laughable if it weren't so tragically misguided—been laid bare in quite this way.

Lown went to see Joseph Garland, editor of *The New England Journal of Medicine*, about publishing the group's findings. And though the *Journal* at the time rarely strayed beyond straight clinical medicine research reports, Garland considered the argument persuasive enough to commit a major series of articles. The articles, which appeared in May 1962, were accompanied by a brief editorial:

> It is no longer a matter of a nation's hiding from the blast or fleeing from it, but of preventing it. This is not to be accomplished unilaterally, by abjection, but by convincing all the participants of the folly of the competition, and showing determined leadership in finding a way out.[52]

After this initial success, Lown gradually bowed out of his leadership role with the Physicians for Social Responsibility. His work on sudden cardiac death had started to gain acceptance, and by the 1970s the device he developed for restarting failing hearts—the defibrillator—was becoming a fixture of emergency rooms throughout the world. Another of his inventions, the Lown cardioverter, was also being widely used to correct heart rhythm abnormalities on a nonemergency basis.[53] During these years Lown devoted more of his attention to his medical research and less to his political activism. But he remembers one beautiful spring day in 1979 when he and his wife were standing outside their house looking at the rolling lawn abloom with new flowers.

> The garden was lush, things were gorgeous, our children were functioning and no longer misbehaving, and yet at that point my wife and I both felt burdened. The reason we were burdened is we had both sort of abrogated social responsibility.[54]

It was time, Lown decided, to get back into the antinuclear movement, which had gained a new urgency in the intervening years with the buildup of enormous nuclear arsenals on either side and the increasing cant about how the "evil empire" of the Soviet Union could not be trusted to keep its finger off the atomic button.

Around this time, Dean Howard Hiatt of the Harvard School of Public Health made an impassioned speech that rallied more people to the antinuclear cause, which he called "the greatest public health hazard [that] now confronts us." As Hiatt put it,

> The lessons of past epidemics have shown us that when no treatment exists and costs are insupportable, all efforts must be directed at prevention. It is essential that society give full attention to the means to prevent nuclear war, lest we face what would otherwise be the last epidemic our civilization will know.[55]

For his part Lown had already established relationships with several Soviet physicians to get his cardioverter accepted internationally. Among them was Eugene Chazov, head of Moscow's esteemed Myasniykov Institute of Cardiology and personal physician to President Leonid Brezhnev and most other members of the Politburo. By 1980 Lown had convinced Chazov to meet with him and two American colleagues on neutral territory, in Geneva, to talk about how to stop the nuclear arms race. Chazov brought along two colleagues as well, and the six doctors settled down to try to agree on strategy. Lown recalls the talks as "very intense":

> We Americans were confrontational. We were so inebriated with the evil of these people that we were saying, in effect, "We nice guys have come to work with you filthy Communists." We didn't understand each other; words we thought were innocent they thought were charged. They said, "Who bombed Hiroshima? Who invaded Vietnam?" We said, "Well, who invaded Afghanistan? What about dissidents?" Finally, I urged that we act as doctors who could face the problem the way doctors had faced smallpox, without asking each other

116

for our political credentials. We decided that we would stick to the nuclear issue to the exclusion of all others and look at it merely from the vantage point of health professionals. The moment we deviated we saw what happened: Over three days, even though we were all the most well-intentioned people, we nearly broke up several times.[56]

Ultimately, good intentions prevailed. At that December meeting Lown and Chazov agreed to cohost a world congress of the fledgling group they were calling International Physicians for the Prevention of Nuclear War (IPPNW). The catch was that the meeting would be held in just three months. The speedy timetable "showed a certain desperation," says Lown, "since we had no staff, no place, no agenda, no participants." But with a personal bank loan of $35,000 as down payment on a conference center in Virginia and with a few months of frantic planning, the IPPNW was able to bring off its first world congress in March 1981—with 75 participants from the Soviet Union, Japan, and several other countries, each of whom had his or her travel expenses paid by the organization, which had managed to win enough foundation grants to cover the costs of the meeting.

The conference received glowing international press coverage and was followed by several more, each one bigger than the last. In 1985, when the group met in Budapest, there were 900 in attendance. Later that year, because of its ability to bring a human and horrific face to the prospect of nuclear war and its ability to help the Soviet Union accept a unilateral moratorium on nuclear testing, the IPPNW was awarded the Nobel Peace Prize.

It should have been a moment of great glory for Lown, who with Chazov was the recipient of the award. But it was tarnished by the ugly reactions of his adopted countrymen (Lown was born in Lithuania and came to the United States as a teenager when his family fled the Nazis in 1935). Americans complained that it was an act of treachery and shame to give a share of the Nobel Peace Prize to Chazov, a high-ranking member of a nation responsible for grotesque human rights violations, such as the imprisonment of dissidents in psychiatric institutions. The IPPNW, critics said, was a pawn of Communist propaganda.

"The Nobel Prize was a nightmare," says Lown with some bitterness. He recalls the noisy joint press conference he and Chazov held in Oslo, just before accepting the award, at which a Soviet correspondent rose to ask a question, obviously troubled that a leading member of Soviet society was being pilloried and shamed. As the man stood up, something unbelievable happened.

> The man had a cardiac arrest. Chazov and I jumped to resuscitate him, the whole world media photographing that, and then the other doctors who were there took over. It was as though God Himself intervened to say, "Look, sudden death can strike anywhere." I went back to the podium and said, "Here the world is about to be struck

Bernard Lown accepting the Nobel Peace Prize

THE
PEOPLE'S
HEALTH

dead, and we doctors are rising up to resuscitate. We do not ask for the politics of the patient, we do not ask his ideology, whether he's kind or unkind to his children. We just have a medical mission." Then I canceled the press conference. But so venomous was the atmosphere that one American correspondent, sitting next to my wife, said, "Those Soviets would stoop to any method to end this press conference."[57]

The correspondent, who was seriously ill and spent several days in a Norwegian hospital, survived. The following day, Bernard Lown—accompanied by his wife, children, and 90-year-old mother—accepted the Nobel Peace Prize. In his speech he reminded those in attendance of the urgency of his mission, as a physician, as a public health advocate, and as a pacifist.[58]

With the dismantling of the Soviet Union in 1989, the immediate threat of nuclear annihilation appeared to ease. But political activists like Lown see a need to remain forever vigilant. Even with the Soviet Union no longer the "evil empire" of our cold war nightmares, he says, the nuclear threat remains with us. Now smaller, more unstable countries, many of them with megalomaniacal leaders, have their metaphoric fingers over the atomic button.

While the world is poised for a public health catastrophe of devastating consequences—one that has so far been averted—we still face daily another type of international holocaust, the death of millions of children from diseases that are easily preventable through a simple vaccine. So while we reel from the implications of nuclear annihilation, we have become inured to the slower but no less certain death caused by lack of the proper injection.

Vaccine Policy and the Politics of Mass Immunization

Mass immunization campaigns, especially for infants and toddlers, are among the most cost-effective strategies in public health. Estimates are that for every dollar spent on immunizing children in the first 18 months of life up to $14 is saved that would otherwise be spent on treatment of infectious disease, days lost from work or school, or premature death.[59]

But despite the clearcut benefits of childhood immunization, Americans are dramatically underprotected. In 1994, as President Bill Clinton's Federal Childhood Immunization Initiative got under way, the CDC estimated that, among children under the age of 2 in the United States, 56 percent were not fully immunized.[60] This is far short of the government's goal of 90 percent immunization by 1996. Poverty partially explains this shortfall; lack of access to services, or lack of money to afford the vaccines themselves, often interferes with a family's willingness or ability to immunize its children. But even relatively affluent parents with good insurance coverage apparently fail to get their children fully immunized. A 1994 study of 1,500 employees of Johnson & Johnson—who were all middle class and all covered by the company's health insurance plan—found that less than half of their two-year-old children were up to date on their vaccinations.[61]

The current system of reimbursement for medical services is partly to blame. In 1994 the CDC conducted a mail survey of 500 pediatricians in New York to find out how they administered vaccines to their young patients, especially those whose parents could not pay for the shots. Half of those responding said they sent some or even all of their patients to a local public health clinic or somewhere else for immunization. Physicians most often sent their patients elsewhere because of "financial hardship for the patient" (88 percent), "lack of coverage of vaccination by private insurance" (54 percent), "discontinuation of free vaccine distribution to doctors by the health department" (45 percent), "vaccine purchase costs" (38 percent), and "insufficient Medicaid reimbursement for vaccination" (30 percent).[62] In most cases the fragmented system of health care for low-income people created "one more opportunity for them to miss their shots."[63]

Children in less developed countries have tended to fare better than some impoverished American children when it comes to immunization. In the mid-1980s the World Health Organization estimated that immunization had saved 800,000 babies worldwide from death by polio, diphtheria, pertussis, tetanus, measles, or tuberculosis. This "major public health gain," the WHO stated, came about through the agency's Expanded Programme on Immunization. In the first decade of the program's existence, 1974 to 1984, immunization rates in the developing world had increased eightfold. The agency's goal when the program began was to achieve a 60 to 70 percent immunization rate by

119

1990.[64] By that time, 16 countries had early childhood immunization rates higher than that of the United States, including Brazil, Bulgaria, Chile, China, Greece, Hungary, Mexico, North Korea, and Romania. "It is remarkable that immunization programs in developing countries have achieved these levels of success in view of the tremendous barriers and situational difficulties that must be overcome," write Gary L. Freed and his colleagues at the University of North Carolina. "For example, during the civil war in El Salvador, two- or three-day ceasefires were called to allow childhood immunization teams safe passage in battle zones."[65]

The failure of a WHO-like commitment to childhood immunization in the United States can be blamed in part on the very success of immunization in an earlier day. We as a nation suffer from a profound collective amnesia about how serious infectious diseases can be. Today's generation of new parents never faced epidemics of smallpox, typhoid, or diphtheria, which were all but wiped out by immunization before the advent of World War II.[66] They cannot even remember the "childhood diseases" that plagued youngsters through the 1960s: polio, measles, mumps, and rubella. That makes it much easier for them to forego the inconvenience and expense of keeping to an immunization routine. At the current recommended levels, the immunization schedule is time consuming and, for the uninsured or underinsured, expensive. The CDC's latest recommendation requires six office visits and 15 different shots or oral doses of vaccine in the first 18 months of a baby's life. Beyond toddlerhood, children need at least four more booster doses—one DTP (diphtheria, tetanus, pertussis), one MMR (measles, mumps, rubella), one dose of oral polio vaccine, and one Td (tetanus and diphtheria)—before they reach the age of 12.[67]

The United States has achieved nearly 97 percent immunization among schoolchildren, since every state mandates a complete set of protective shots for all children entering school. But a nation of well-protected 5 year olds is still vulnerable to preventable childhood infections. As Freed and his coauthors point out:

> Immunization at school entry does not prevent the life-threatening diseases of infancy for which most of these vaccines exist. Epidemics of vaccine-preventable diseases have recently occurred in many cities throughout the United States, including Houston, Los Angeles, and New York City, establishing our national problem of poor early childhood immunization rates as a significant public health concern.[68]

When children come down with easily preventable childhood diseases for which vaccines are available, that is a sign of a health care system in trouble. Significantly, the rates of both measles and rubella went up sharply in the early 1990s, despite the availability of safe and effective vaccines that all children should have received by the age of 15 months. In 1990 more than 27,000 cases of measles were reported in the United States, a 52 percent increase over the previous year and a nearly 20-fold increase over the record low set in 1983. Similarly, the incidence of rubella increased by 500 percent between 1988 and 1991—with the annual number of cases of congenital rubella syndrome, a serious constellation of birth defects that had all but disappeared in the late 1980s, increasing during that time from two to 24. "Such events," write Freed and his colleagues, "should serve as a harbinger of the direction of children's immunization status and as a call to action to address deficiencies."[69]

One massive immunization campaign that sullied the reputation of other such campaigns for years afterward was the ill-fated swine flu episode—a moment in time still referred to by public health historians as the "swine flu fiasco." The large-scale effort in 1976 to administer a vaccine to "every man, woman, and child in the United States" (in the words of President Gerald Ford) demonstrated that the federal government, in partnership with the pharmaceutical industry, was capable of manufacturing and distributing millions of doses of high-quality vaccine in a matter of months. But because the dreaded pandemic never materialized, it seemed in hindsight that 40 million Americans were vaccinated for nothing. Indeed, the only real illness connected to the swine flu adventure was actually caused by the vaccine: about 1,000 people developed Guillain-Barre syndrome, a serious paralytic disease that could be traced directly to an immunological response to the inoculation.[70]

The story began in February 1976, on Friday the thirteenth. Reports came in to the CDC of an influenza outbreak among a few dozen men at Fort Dix, a boot camp in New Jersey. One recruit, 19-year-old Private David Lewis, died. Of 19 specimens examined for typing, five, including the one from Pvt. Lewis, looked like the same virus that probably caused the 1918 flu pandemic, when a highly virulent strain killed 20 million to 40 million people worldwide. Many flu experts believed that the next pandemic had arrived but that modern technology and rapid vaccine production could help avert hundreds of thousands of deaths.

In early March, David Sencer, director of the CDC and a 1958 graduate of the Harvard School of Public Health, sought the advice of a panel of immunization experts. Among them was Reuel Stallones, dean of the public health school at the University of Texas, who later recalled feeling that the right decision might finally be a way to give public health credit for helping save lives—credit it had long deserved and failed to receive.

> It was an opportunity to strike a blow for epidemiology in the interest of humanity. The rewards have gone overwhelmingly to molecular biology which doesn't do much for humanity. Epidemiology ranks low in the hierarchy—in the pecking order, the rewards system. Yet it holds the key to reducing lots of human suffering.[71]

Like many of his advisers, Sencer believed that, even if swine flu never materialized, it was better to err on the side of caution. In a much-cited "action memo," Sencer pushed for universal immunization in the face of uncertainty. "The Administration can tolerate unnecessary health expenditures," he wrote, "better than unnecessary death and illness."[72] In this he was echoing the sentiments of one of the virologists on his expert panel, Edwin Kilbourne of Mt. Sinai Medical School in New York. Kilbourne, an early and consistent advocate of mass immunization, was fond of saying "better a vaccine without an epidemic than an epidemic without a vaccine."[73]

But in hindsight scientists involved in the decision making have learned that major variants of influenza virus in a few individuals do not necessarily signal the start of a new pandemic. "Early detection of a new virus may not be adequate evidence on which to undertake mass immunization," Kilbourne wrote three years later in an article entitled "Swine Flu: The Virus That Vanished." "But it is, I believe, a signal at least to produce vaccine to hold in readiness."[74]

The CDC opted to not stockpile vaccine; instead, it placed the country on active alert. By October 1976 a massive system of free immunization was in place nationwide, and 40 million Americans rolled up their sleeves in high school gyms, city halls, and shopping malls across the country. The system worked smoothly for about six weeks, until reports started trickling in from Minnesota, Alabama, and other states of a rare paralytic condition, Guillain-Barre syndrome, in people who had received the swine flu vaccine. By the middle of December, government epidemiologists established a cause-and-effect connection between the vaccine and Guillain-Barre. On December 16, with the concurrence of President Ford, U.S. Health, Education, and Welfare Assistant Secretary for Health Theodore Cooper suspended the swine flu immunization program "in the interest of safety of the public, in the interest of credibility, and in the interest of the practice of good medicine."[75]

The "swine flu fiasco" turned out to have some long-lasting political repercussions. CDC Director Sencer resigned; some say even President Ford's defeat in the November elections, one month after vaccinations began and one month before they were aborted, could be traced to his part in pushing an unpopular mass vaccination program.

The swine flu vaccine was the most public forum in which vaccine policy was debated and revised. But the federal government, through a CDC panel known as the Advisory Committee for Immunization Practices (ACIP), continually reviews vaccines that have been approved by the Food and Drug Administration, new and old ones alike, to decide which to recommend for use by a small segment of the population (such as the typhoid vaccine) or for wider use. Ever since the 1970s, for instance, the ACIP has revisited the question of which polio vaccine is better for American children—the oral Sabin vaccine now in use, which uses a live weakened virus, or the original Salk injectable vaccine, which uses a killed virus.

"Every year there have been a handful of cases of paralytic polio in the United States that are related to the vaccine strain," says Mary Wilson of the Harvard School of Public Health, who was on the ACIP from 1988 to 1994. "At a time when there are no cases at all occurring in the United States because of natural transmission of the wild virus, some people have argued that we should not accept any cases of paralysis or deaths from the vaccine strain." On the other hand, the reason for the lack of wild-type polio cases could in part be the generalized use of the live-virus vaccine, which conveys immunity even to nonimmunized adults and children who are exposed to the vaccine virus when it is shed in the feces of a person who was vaccinated. "There have been some outbreaks of polio in countries where they have used only the killed vaccine," Wilson points out. "So neither one is perfect."[76] In late 1995 the CDC proposed a compromise solution that would utilize both vaccines, such as a killed virus for the first two doses and a live virus for the third. This arrangement still confers some community-based immunity to nonimmunized contacts but at less potential risk to the child getting the vaccine. Whether the proposed compromise represents progress remains debatable.

New vaccines are continually being developed and licensed, and the government must decide how and whether to recommend their inclusion in the general childhood immunization schedule. The government must also decide which recommendations to issue for adult vaccines, such as those against pneumococcus (which causes pneumonia) and influenza. Recently, an old tuberculosis vaccine known as BCG was considered, at least for certain high-risk groups, but there is still much resistance to using it because the vaccine seems to be only 50 percent effective and, as a live vaccine, it occasionally causes disease in immunocompromised individuals. In addition, people who receive BCG may subsequently test positive to a tuberculin skin test, which is currently the only method of screening for TB in the general population. "There are a lot of passions about BCG," Wilson observes. "People tend to have very strong feelings and not all are supported by the facts that are available."[77]

Wilson and her colleagues at the Harvard School of Public Health conducted a meta-analysis to determine the efficacy of BCG and found that it is nowhere near the 90 or 95 percent that most Americans have come to expect in widely used vaccines. As she says:

> We need a better vaccine. We also need a better diagnostic test for tuberculosis. If we had more specific tests, like a blood test instead of the very, very crude tuberculin skin test we're using now, it would remove that issue from the debate over BCG. But still, I think people in this country are not going to accept BCG for general use. People like vaccines that have no side effects and very high efficacy.[78]

Vaccine policy brings us to the peculiar intersection of social policy and individual decision making. Government authorities can debate endlessly about the relative safety of killed versus live polio vaccine, but unless mothers bring their babies in for immunization at 2, 4, 6, 12, 15, and 18 months of age, the debate is wasted. In Chapter 5, we turn to the next critical piece in the puzzle of how to preserve the public's health: efforts to provide people with guidance to make the right health choices for themselves and their loved ones every day of their lives.

Before beginning our discussion of individual behavior and its contribution to public health, we should take a moment to remember that even the most personal action still is made in a social context. Dorothy Nelkin of New York University and Sander Gilman of Cornell are among those who believe that too great an emphasis on individual responsibility absolves society at large from many public health problems that are really societal in origin. If we insist that lung cancer, for instance, is caused by cigarette smoking, we ignore the contributions to cancer of environmental toxins and, more importantly, the social context in which smoking occurs: cigarette advertisements that appeal to preteens and teenagers, the ready availability of cigarettes to minors, the relatively low cost of cigarettes through price supports and lack of excise taxes, manipulations of the nicotine content of cigarettes to increase their addictive properties, and the decades-long consensus that smoking was acceptable behavior. These social factors help lead to an individual behavior—smoking—but they are not really within an individual's control.

The intimate connection between individual choices and social constraints was outlined in *Healthy People 2000*, prepared by the U.S. Public Health Service in 1990 to set out health goals for the turn of the century. Many health goals "with a prominent behavioral component" were not the domain of the individual alone, the document emphasizes, but required "strategies that address the social context that gives shape to personal attitudes and choices."[79] The report made clear that even the decisions that individuals make about their day-to-day lives are made in reaction to a broader—and sometimes none-too-healthful—social and political environment.

The decisions people make about how to live their lives also reverberate in the society in which they live. One way to see this is by looking at the increase in the consumption of low-fat yogurt, which more people choose as evidence accumulates about the health benefits of a low-fat diet. But Walter Willett of the Harvard School of Public Health points out that every dietary choice we make, even if it is a healthy one, ultimately affects the nutritional strategies of society as a whole. Yes, choosing low-fat yogurt might be a healthy step in terms of one individual's daily diet, but when evaluated in the context of the population at large, Willett says, low-fat yogurt is likely to have little impact on the public's health.

> Once the milk gets produced, somebody will eat the fat. If you eat low-fat yogurt, the fat that is extracted is not used for jet fuel or to power automobiles. It goes somewhere else in the food supply.[80]

Willett likens excess animal fat to hazardous waste; it keeps getting passed from one community to another, eventually trickling down to the poorest and most powerless. Much of the fat that is removed from yogurt and whole milk, he says, finds its way into school lunches or cheese and butter giveaways—the kind of food that children and poor people eat.

This perspective is important to bear in mind as we review in the next chapter the behavioral contributors to major public health disasters. While it is easy to see the biggest killers of the late twentieth century as wholly the responsibility of individual people making individual decisions, the context in which they make those decisions is less obvious but no less important.

Notes

1. Lisa F. Berkman and Lester Breslow, *Health and Ways of Living: The Alameda County Study*. New York: Oxford University Press, 1983, p. 28.

2. Lisa F. Berkman and S. Leonard Syme, "Social networks, host resistance, and mortality: a nine-year follow-up study of Alameda County residents," *American Journal of Epidemiology*, 1979, vol. 109, no. 2, p. 188.

3. Ibid., pp. 190–191.

4. *Report of a General Plan for the Promotion of Public and Personal Health, Devised, Prepared, and Recommended by the Commissioners Appointed Under a Resolve of the Legislature of the State.* Boston: Dutton & Wentworth, 1850. Reprinted by Harvard University Press, Cambridge, Mass., 1948.

5. Berkman and Breslow, *Health and Ways of Living*, p. 18.

6. Benjamin Paul, *Health, Culture, and Community: Case Studies of Public Reactions to Health Programs.* New York: Russell Sage Foundation, 1955.

7. Alexander H. Leighton, *My Name Is Legion: Foundation for a Theory of Man in Relation to Culture.* New York: Basic Books, 1959.

8. Elizabeth Fee and Barbara Rosenkrantz, "Professional education for public health in the United States." In E. Fee and R. H. Acheson, eds., *A History of Education in Public Health: Health That Mocks the Doctors' Rules,* Oxford and New York: Oxford University Press, 1991, p. 249.

9. John Duffy, *The Sanitarians: A History of American Public Health.* Urbana: University of Illinois Press, 1990, p. 274.

10. Sol Levine, personal interview, Sept. 1995. Levine expands on these ideas in the introductory chapter he wrote with Benjamin C. Amick, Alvin R. Tarlov, and Diana Chapman Walsh to their book, *Society and Health* (New York: Oxford University Press, 1995).

11. Bruce Kennedy, Ichiro Kawachi, and Deborah Prothrow-Stith, "Income distribution and mortality: cross-sectional ecological study of the Robin Hood Index in the United States," *British Medical Journal,* April 20, 1996, vol. 312, pp. 1004–1007.

12. William Foege, personal interview, March 1994.

13. Ibid.

14. John S. Billings, "Public health and municipal government," *Annals of the American Academy of Political and Social Science* (Suppl.), Feb. 1891, p. 6, cited in Dorothy Nelkin and Sander Gilman, "Placing blame for devastating disease," in Arien Mack, ed., *In Time of Plague: The History and Social Consequences of Lethal Epidemic Disease,* New York: New York University Press, 1991, p. 48.

15. H. Jack Geiger, "Community health center: health care as an instrument of social change." In V. W. Sidel and R. Sidel, eds., *Reforming Medicine: Lessons of the Last Quarter Century,* New York: Pantheon Books, 1984, pp. 15–16.

16. Alonzo S. Yerby, "The disadvantaged and health care," *American Journal of Public Health,* 1966, vol. 56, pp. 6–7.

17. U.S. National Center for Health Statistics, *Vital Statistics of the United States: Death Rates for Accidents and Violence, 1980–1991,* 1994, p. 100.

18. George Rosen, *A History of Public Health, Expanded Edition.* Baltimore: Johns Hopkins University Press, 1993, p. 389.

19. Duffy, *The Sanitarians,* p. 2.

20. Joycelyn Elders, personal communication, Oct. 18, 1994.

21. Camara Jones, personal interview, Oct. 1995.

22. Ibid.

23. As historian George Rosen explains in *A History of Public Health, Expanded Edition* (Baltimore: Johns Hopkins University Press, 1993, pp. 318–319), "Housing is important, for example, because overcrowding favors the spread of respiratory infections and lack of adequate washing facilities increases gastrointestinal infection. The level of infant mortality varies as well with the availability of medical care and proper knowledge of infant nutrition."

24. After a steadily decreasing death rate among newborns in the first half of this century, the United States lost its footing. Today it has among the worst infant mortality rates in the industrialized world. According to David Kotelchuck, editor of *Prognosis Negative: Crisis in the Health Care System* (New York: Vintage Books, 1976, p. 6), the United States in 1955 had the eighth-lowest infant mortality rate among the world's 20 most industrialized nations, with a death rate of 26 newborns per 1,000 live births. But by 1973, even though the death rate had declined to 18 deaths per 1,000 births, the United States had not kept up with progress in the other industrialized countries, and its ranking slipped to fifteenth out of 20, with only Ireland, West Germany, Czechoslovakia, Austria, and the USSR faring worse.

25. Steven Gortmaker, principal investigator, *A Proposal for the Creation of a Community-Based Health Intervention to Reduce Infant Mortality and Improve Birth (and Life) Outcomes.* Boston: Harvard School of Public Health, Feb. 1987, p. 13.

26. Marie McCormick, personal interview, Aug. 1995.

27. Kristen J. Mertz, Artist L. Parker, and George J. Halpin, "Pregnancy-related mortality in New Jersey, 1975 to 1989," *American Journal of Public Health,* 1992, vol. 82, p. 1085.

28. Howard Hiatt, "The physician and national security," *The New England Journal of Medicine,* 1982, vol. 307, p. 1142.

29. Felton Earls, personal interview, May 1995.

30. Ibid.

31. Sasha Cavender, "Teen treatment: doctor prescribes an antidote to America's violence," *The Chicago Tribune,* Dec. 19, 1993.

32. Valerie Adler, "The roots of violence," *Harvard Public Health Review,* Winter 1991, p. 5.

33. One of the ways the CDC used to focus attention on injury as a public health problem was devised by William Foege, CDC director and an alumnus of the Harvard School of Public Health. He introduced the concept of "years of life lost" as a standard measure of the burden of illness, thus elevating the visibility of injury and its cost to society.

34. Deborah Prothrow-Stith with Michaele Weissman, *Deadly Consequences.* New York: HarperCollins, 1991, p. 135.

35. "Homicides among 15-to-19-year-old males—United States, 1963–1991," *Morbidity and Mortality Weekly Report,* 1994, vol. 43, pp. 728–730.

36. Gary Taubes, "Violence epidemiologists test the hazards of gun ownership," *Science,* 1992, vol. 258, p. 215.

37. Ibid., p. 214. The editorial, written by Jim Mercy and Vernon Houk of the CDC, appeared in *The New England Journal of Medicine* in Nov. 1988.

38. Felton Earls, "Not fear, nor quarantine, but science: preparation for a decade of research to advance knowledge about causes and control of violence in youths," *Journal of Adolescent Health*, 1991, vol. 12, p. 622.

39. Deborah Prothrow-Stith, "Violence: a complex health hazard," *The Chicago Tribune*, Dec. 28, 1993.

40. Peter Applebome, "CDC's new chief worries as much about bullets as about bacteria," *The New York Times*, Sept. 26, 1993, p. 7.

41. Ibid.

42. Dorothy Cheek, "America's schools experience escalating violence among students," *Nation's Cities Weekly*, Feb. 7, 1994.

43. Mukiya Adams's essay was reprinted in Prothrow-Stith, *Deadly Consequences*, p. 80.

44. DeNeen L. Brown, "Afraid to die, afraid to live: as a survival tactic, many teens retreat from social activity," *The Washington Post*, Jan. 16, 1995, p. A8.

45. Prothrow-Stith, *Deadly Consequences*, p. 22.

46. In 1987 the U.S. Public Health Service developed a national set of goals for the turn of the century, called *Healthy People 2000*, that included the aim of reducing the rates of adolescent homicide. At the time, the homicide rate for black males between the ages of 15 and 34 was 10 times the rate of the general population: 91 per 100,000 young black males, compared to 8.5 per 100,000 overall. Although the goal was to reduce the rate by 20 percent, to 72 per 100,000, a midcourse review of the plan published in 1995 found that the situation had gotten worse instead of better. At the time of the update, black males were being killed at a rate of 134 per 100,000, and the population at large was 10.5 per 100,000—both far short of the year 2000 goal of 72 and 7.1 per 100,000, respectively (Sally Squires, "Report on prevention project is mixed bag," *The Washington Post*, Health section, April 18, 1995, p. 7).

47. Prothrow-Stith, *Deadly Consequences*, p. 139.

48. Bernard Lown, personal interview, May 1994.

49. In the years since, the Physicians for Social Responsibility broadened its agenda to include a host of other issues, including campaigns against torture, environmental hazards, low-level radiation, and renegade nuclear weapons.

50. These figures, according to Lown, were derived from the Atomic Energy Commission, which estimated that in the event of an all-out nuclear war in 1961 the United States was likely to receive a total of 1,000 megatons dropped on various strategic cities, including 10 megatons on Boston.

51. Lown, personal interview.

52. Joseph Garland, "—Earthquake, wind and fire," *The New England Journal of Medicine*, 1962, vol. 266, p. 1174.

53. Bill Dersiewicz, "Bernard Lown speaks from the heart," *Harvard Public Health Review*, Winter 1990, p. 5. Lown also introduced the now-accepted use of lidocaine to treat life-threatening arrhythmias in heart attack patients.

54. Lown, personal interview.

55. "Dean Hiatt, alumni Frechette and Geiger active in anti-nuclear war movement," *Harvard Public Health Alumni Bulletin*, Fall 1982, vol. 38, no. 1, p. 12.

56. Lown, personal interview.

57. Ibid.

58. In his acceptance speech, reprinted as "A prescription for hope" in *The New England Journal of Medicine* (1986, vol. 314, pp. 985–987), Lown said in part, "We have resisted being sidetracked to other issues, no matter how morally lofty. Combating the nuclear threat has been our exclusive preoccupation, since we are dedicated to the proposition that to ensure the conditions of life, we must prevent the conditions of death. Ultimately, we believe people must come to terms with the fact that the struggle is not between different national destinies—between opposing ideologies—but between catas-

Confronting the Social Environment

trophe and survival. . . . If we are to succeed, this vision must possess millions of people, we must convince each generation that they are but transient passengers on this planet earth. It does not belong to them. They are not free to doom generations yet unborn. They are not at liberty to erase humanity's past or dim its future. Only life itself can lay claim to sacred continuity."

59. Chester A. Robinson, Stephen J. Sepe, and Kimi F. Y. Lin, "The president's child immunization initiative—a summary of the problem and the response," *Public Health Reports,* 1993, vol. 108, pp. 419–420. Immunization experts point out, though, that not every vaccine is cost effective. As Mary Wilson of the Harvard School of Public Health says, "People have the sense that because immunization is preventive medicine it is inherently cost effective. Some of the early vaccines, particularly those that were not particularly expensive and that protected against diseases that were very widespread, have been shown in analyses to save more than they cost. But many other vaccines, especially those that are used against infections that are rare, are extremely expensive when you calculate the cost per case averted" (Mary Wilson, personal interview, Oct. 1995).

60. Don White, "Clinton administration asks HMOs to take the lead in national initiative to boost childhood immunization rates," news release, Group Health Association of America, May 9, 1994.

61. Elyse Tanouye, "Study shows even the well-insured fail to meet immunization-rate goals," *The Wall Street Journal,* Feb. 16, 1994, p. B5.

62. Donald M. Berwick, "Vaccination: doctors often pass the buck," *Journal Watch,* Feb. 15, 1995, vol. 15, p. 35. In his article, Berwick cites the CDC's report, "Physician vaccination referral practices and vaccines for children—New York, 1994," *Morbidity and Mortality Weekly Report,* 1995, vol. 44, pp. 3–6.

63. Berwick, "Vaccination."

64. "Immunizations save 800,000 infant lives yearly," *World Health,* Nov. 1985, pp. 30–31.

65. Gary L. Freed, W. Clayton Bordley, and Gordon H. DeFriese, "Childhood immunization programs: an analysis of policy issues," *The Milbank Quarterly,* 1993, vol. 71, p. 68.

66. Harry F. Dowling, *Fighting Infection: Conquests of the Twentieth Century.* Cambridge, Mass.: Harvard University Press, 1977, pp. 23–35.

67. Centers for Disease Control, "Recommended childhood immunization schedule—United States, January 1995," *Journal of the American Medical Association,* 1995, vol. 273, p. 693.

68. Freed, Bordley, and DeFriese, "Childhood immunization programs," pp. 66–67.

69. Ibid., p. 67.

70. Robin Marantz Henig, *A Dancing Matrix: How Science Confronts Emerging Viruses.* New York: Vintage, 1994, p. 178.

71. In 1978, U.S. Department of Health, Education, and Welfare Secretary Joseph Califano asked Richard Neustadt of Harvard University and Harvey Fineberg of the Harvard School of Public Health to analyze the events of early 1976 through the following March. "The swine flu program is now overwhelmingly recalled as a 'fiasco,' a 'disaster,' or a 'tragedy,' " they wrote in the introduction to their report, *The Swine Flu Affair: Decision-Making on a Slippery Disease,* published by HEW in 1978. The failed attempt to immunize every American against a new variant of influenza, they wrote on page 12, "was and is a trauma to the government officials most involved and to their scientific advisers. A year and more later, cheeks flush, brows furrow, voices crack." Stallones's quote appears in this book on p. 12. An expanded edition was published as *The Epidemic That Never Was: Policy-Making and the Swine Flu Affair,* New York: Vintage Books, 1982.

72. Henig, *A Dancing Matrix,* p. 178.

73. Ibid.

74. Edwin D. Kilbourne, "Swine flu: the virus that vanished," *Human Nature,* March 1979, p. 73.

75. Neustadt and Fineberg, *The Swine Flu Affair,* p. 70.

76. Mary Wilson, personal interview, Oct. 1995.

77. Ibid.

78. Ibid.

79. J. Michael McGinnis and Philip R. Lee, "*Healthy People 2000* at mid decade," *Journal of the American Medical Association*, 1995, vol. 273, p. 1123.

80. Peter Wehrwein, "Food fight," *Harvard Public Health Review*, Fall 1994, vol. 6, p. 39.

*Confronting
the Social
Environment*

 **5**

Providing Guidance for
Individual Behavior

Imagine a man named Patrick O'Malley, who came to America from Ireland in the 1930s. He was 22 years old and looking for the good life. Patrick settled in Boston, where he had many aunts and uncles. Job opportunities were limited, but young Patrick was able to find work as a police officer. When he first arrived in Boston, his cousin introduced him to Katie Cullen, a pretty Irish girl who was working as a housekeeper at a home in Brookline. Within a year, Patrick and Katie were married, and together they raised four sons and a daughter. Patrick performed his job well and was steadily promoted in the police department and in the 1960s achieved his highest rank, that of lieutenant.

Now imagine Patrick's older brother, Sean, who stayed behind in Dublin. Sean, too, married early and had five children, two boys and three girls. He never made as much money as his brother in America, but he never needed as much either. All he needed was enough to keep his bicycle in shape to get him to work as a housepainter and to get him to the pub before closing time.

The O'Malley brothers are inventions, but just picturing them allows one to make conjectures about whole facets of their lives: how they eat, whom they socialize with, what their communities are like, and what the future holds for their children. Some of those differences have definite health implications, as demonstrated in the landmark study conducted by the Harvard School of Public Health in collaboration with the School of Medicine of Dublin's Trinity College known as the Boston-Ireland Brothers Study.

The study, originally conducted in the 1960s, involved some 500 pairs of brothers in all. Within each pair, one brother lived in Dublin and the other in Boston. The Boston brothers had all lived in America for at least 10 years at the beginning of the study, and the brother pairs were all relatively close in age (no more than 10 years apart). The Boston-Ireland Brothers Study results, reported in 1970, were among the first to elucidate the health implications of one's social surroundings. (A similar study comparing siblings in Israel and the United States, by Ascher Segall at the Harvard School of Public Health, was conducted at about the same time.) Most things about the Boston-Irish brothers were similar—their ages, genetic backgrounds, individual disease risks, even their incomes. The differences in their diets were small and without much significance in terms of known health consequences: the Irish brothers took in more calories, complex carbohydrates, magnesium, and (probably because they drank so much tea) fluoride.

Yet despite their similar backgrounds, incomes, and diets, the two sets of brothers had significantly different rates of heart disease. The brothers who had emigrated had significantly more heart disease than the brothers who had stayed in Ireland. They had more atherosclerosis, weighed more, had higher proportions of body fat, and had higher numbers of abnormal electrocardiograms. The explanation for these differences seemed to lie in the men's different levels of physical activity. With similar genetic endowments, the brothers had similar serum cholesterol and blood pressure levels. But because of the cultures in which they lived, the brothers in Ireland engaged in a far greater amount of physical activity—and that seemed to be what saved them from coronary heart disease. These findings made it clear that your everyday habits can be as important to your health status as the genes you start out with, the germs you are exposed to, or the food you eat.[1]

By the postwar years, the American public was dying of distinctly different diseases than had plagued it in the first half of the century. In 1900 the leading causes of death were infectious diseases: tuberculosis, pneumonia, diarrheal diseases, and enteritis. But by 1946 the tide had turned, and the two leading causes of death were chronic illnesses: heart disease and cancer. A new category, injuries, held the number three slot.

When taken together, then, the most common causes of death in midcentury—and in all the years since—have had in common one salient characteristic: their relationship to some aspects of individual behavior. Heart disease, as the research of the past 50 years has shown, is in large measure the result of smoking, poor diet, lack of exercise, and failure to treat high blood pressure; some forms of cancer can be traced as well to smoking, diet, and other lifestyle choices; and injuries and other traumatic deaths are the direct outgrowth of risky behavior in association with lethal weapons like automobiles and guns.

"Approximately 50 percent of deaths in people under 75 are due to personal behaviors which can be modified," the U.S. Public Health Service announced in 1994. This finding was based on a number of studies, including one released the previous year that found that one-half of the 2.1 million deaths in 1990 were the result of tobacco smoking, alcohol use, poor diet, lack of exercise, firearms, automobiles, drug abuse, and similar "lifestyle risk factors."[2]

This chapter looks at some of the things people do that help determine their individual health—which, when considered in the aggregate, helps determine the health of whole populations. This investigation brings us to some of the most familiar, yet at the same time some of the most controversial, issues in the field of public health, including:

- risk factors for heart disease, cancer, and other chronic illnesses, as elucidated in large-scale population studies such as the Framingham Heart Study and the Harvard Nurses' Health Study;
- cigarette smoking, the single biggest contributor to preventable cancer and heart disease;
- nutrition, including often-conflicting advice about the contribution to health and illness of food components such as fat, dietary cholesterol, alcohol, and total calories;
- risky driving behavior, especially in terms of drunk driving, and the effort to reduce traffic deaths and promote the notion of the "designated driver"; and
- AIDS, arguably the most devastating disease of the late twentieth century, which is passed through the things people do—in terms of sexual practices and drug abuse—and can be contained, if not eliminated, through behavioral interventions.

In addition, we will look at the science of risk assessment, which evaluates the true nature of a particular risky behavior in the context of comparable risks. First, we turn to the way that risk is conceptualized most frequently, in the form of the "risk factor"—a term that did not exist before 1950, when it was popularized in the familiar Framingham Heart Study.

Risk Factors and Populations: The Framingham Heart and Harvard Nurses' Health Studies

By midcentury, experts did not know much about what caused heart disease; they only knew that it was on the rise. Death rates had increased from 165 per 100,000 adults in 1900 to 200 per 100,000 in 1920 and to 250 per 100,000 in 1930.[3] But no one had yet figured out who was getting heart disease and how they differed from those who did not. "Most of the cases in the early part of the century were due to rheumatic heart disease, not coronary heart disease," Walter Willett of the Harvard School of Public Health explains. "Death from coronary heart disease was not even recognized in the first decade of this century but by 1950 had become the number 1 cause of death."[4] Americans were just about as helpless then in the face of heart disease as they are today in the face of many cancers.

For evidence of how little was known in the 1940s about predisposing conditions to cardiovascular disease, consider President Franklin D. Roosevelt's history of high blood pressure. Today we know that hypertension is one of the most significant risk factors for heart disease, and most doctors start to treat it in adults when it creeps above the "borderline normal" range of 140/90. But when Roosevelt was at Yalta in February 1945, his blood pressure was an extremely high 260/150. This was the highest point of a long progression of ever-increasing blood pressure readings, according to Roosevelt's medical records: 188/105 in 1941, 226/118 just before D-Day in 1944. On the morning he died in April 1945, his blood pressure suddenly zoomed to a catastrophic 300/190. Yet despite his medical history, Roosevelt's doctors expressed shock at his death. Even though cerebral hemorrhage, the immediate cause of death, is now known to be a common complication of hypertension, Roosevelt's doctors told reporters that the president's stroke "came out of the clear sky."[5]

Not only were leading cardiologists of the 1940s unaware of the dangers of high blood pressure, but many of them actually believed it might be good for you. The conventional wisdom of the day was that hypertension was compensatory in individuals who, like Roosevelt, had severe atherosclerosis that narrowed their blood vessels. Even as eminent a cardiologist as Paul Dudley White of Harvard believed that a higher pressure might be necessary to force a sufficient volume of blood through constricted routes. As White wrote in the 1937 edition of his classic textbook on heart disease:

> The treatment of the hypertension itself is a difficult and almost hopeless task in the present state of our knowledge, and in fact for aught we know . . . the hypertension may be an important compensatory mechanism which should not be tampered with.[6]

White was among a group of prominent physicians in the Boston area who in 1948 recognized that "the present state of our knowledge" was insufficient for dealing with the growing public health threat posed by coronary heart disease. To expand on that knowledge, they designed a huge prospective longitudinal study to see who developed heart disease and why.

There was some precedent for large-scale longitudinal studies. In 1930, for instance, a Harvard instructor began what turned out to be one of the classic population studies of human growth with little more than a pair of calipers. Harold Stuart, a young pediatrician at Children's Hospital in Boston, borrowed the calipers from Earnest Albert Hooton, a Harvard anthropologist who made a career of measuring the heads and bottoms of a wide range of creatures: apes, gorillas, criminals, college students. Hooton was to become renowned in the 1940s for going into New York's Grand Central Station and measuring 10,000 fannies, as part of an assignment to design the ideal railroad seat.[7]

Isabelle Valadian

With those calipers and other anthropometric instruments, Stuart at first intended to follow the growth of infants from before birth (by measuring their mothers and recording their mothers' diets) until age 2. "It soon became obvious," he later recalled, "that to really study growth and development properly, it was necessary to start with infants and follow them continuously."[8] Eventually, Stuart's study—which was carried on after his death by Isabelle Valadian and Fredrick Stare of the Harvard School of Public Health—involved close follow-up of 300 youngsters from the Roxbury section of Boston, from before birth through the age of 21. As a result of the Harvard Growth Study, physicians were able to advise parents about how their children's growth rates compared to the average. The most dramatic use of the growth curves was as an early warning system for spotting malnourished children—some of whom, on casual observation, might appear to be growing normally.

With this as background, the heart disease experts who assembled in 1948 knew that the only way to understand the origins and natural history of a condition was to follow a large group of individuals for a long period of time. Because their group included people from the Harvard Medical School and Boston and Tufts universities, the investigators chose to focus on residents of the nearby community of Framingham, Massachusetts. Their study—still alive and well nearly 50 years later—came to be known as the Framingham Heart Study.

Framingham was chosen for several reasons. It was accessible: a straight shot down the Massachusetts Turnpike, 22 miles from Boston. It was reasonably homogeneous: most of the residents were white, of Italian or Irish ancestry with a smattering of WASPs and Jews.[9] And, most important, it was stable: in the first 40 years of the Framingham Heart Study, even those few residents who moved away (usually to Florida for retirement) kept coming back to be examined, and the investigators managed to keep tabs on a full 97 percent of the original study subjects.[10]

The original cohort consisted of about 5,200 healthy people (about 2,600 men and 2,600 women) between the ages of 30 and 62. The plan was to put each subject through an hour-long medical examination every two years and to have each submit to a lengthy questionnaire about eating, exercising, and other daily activities. Because they did not know which factors would turn out to be important, the scientists included everything they could think of: blood pressure, lung capacity, body fat, height and weight, diet, drinking habits, personality traits, family situa-

AGE 45–54

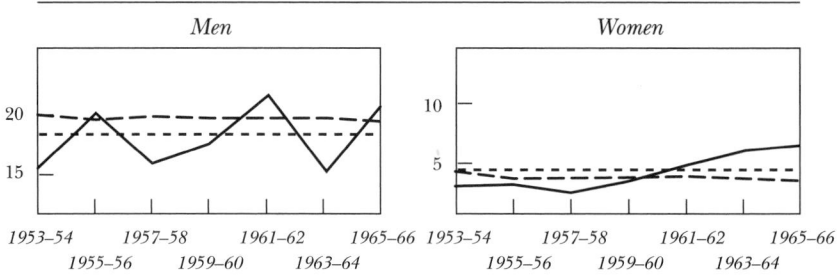

AGE 55–64

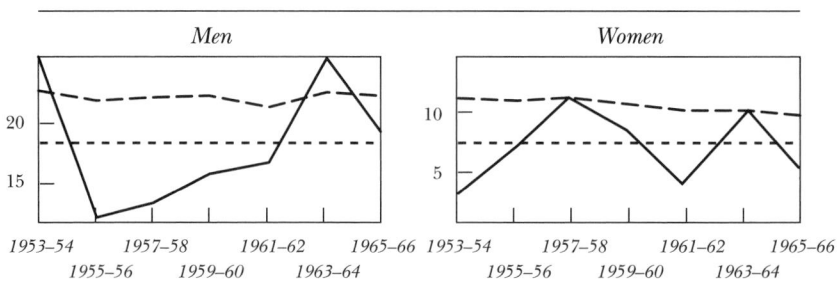

AGE 65–69

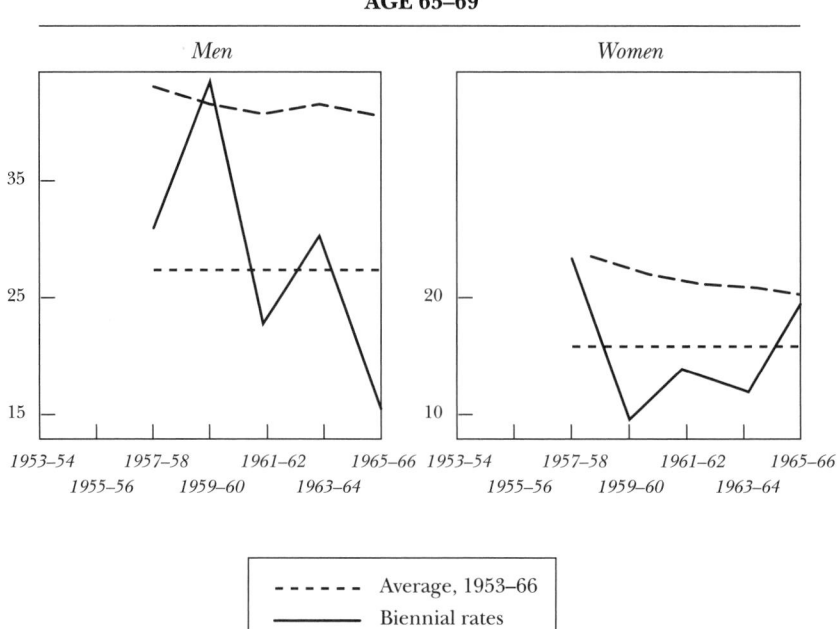

Death rates by age and sex, Figure 1-1 from The Framingham Study, 1968.

136

tions, everything from the sublime to the slightly ridiculous. One of the early advisers to Framingham, a prominent Boston cardiologist, expected exercise to turn out to be important in terms of heart disease risk, but his reasoning was backward: he believed that each person had a lifetime allotment of heartbeats and that active people would use up theirs more quickly and therefore die younger.[11]

The Framingham investigators followed each subject until death and then found out how he or she died. Then came the hardest part: trying to find an association between manner of death and something about the way that person lived. After the study had been going on for about 10 years, those first associations became possible. Thus was born the notion of "risk factors": the idea that identifiable elements of an individual's background, physical condition, or daily habits made someone more likely to develop a particular disease.

"We were the first ones to use the term," says Thomas Dawber, who directed the Framingham study from 1950 to 1965.

> We decided way at the beginning that what we were studying were factors that increased your risk of getting disease. It was a statistical concept, really; I guess most of this came from the statisticians we had in the organization.[12]

Among the risk factors for heart disease elucidated through the data from Framingham were sedentary lifestyle; cigarette smoking; high blood pressure (contrary to earlier belief, the Framingham researchers also found that the systolic reading was at least as important a predictor of heart disease as the diastolic); high total serum cholesterol, especially if accompanied by a low level of high-density lipoprotein; and being male.[13]

The Framingham data have proved to be a precious stockpile for investigators working in many other fields of study across the country. The existence of all that background population data, as well as stockpiles of frozen blood collected over the years, continues to make the Framingham collection a valuable resource for scientists who today want to know the distribution of certain variables within a normal population of healthy white Americans. Dawber remembers that in the 1950s, when many communities resisted water fluoridation in the mistaken belief that it would increase cancer rates, Framingham data helped put those fears to rest. As he recalls:

> The National Institute of Dental Research wanted us to compare a couple of counties in Texas that had high levels of fluoride in the water with Framingham, which has a very low level of fluoride in the water. We showed conclusively that the high rates of fluoride in the water in Texas did not increase the cancer rate.[14]

The Framingham Heart Study has proved so successful that in 1971 investigators were able to embark on a sort of Son of Framingham.

Officially known as the Framingham Offspring Study, this second-generation longitudinal study involves the sons and daughters of the 1,600 married couples in the original cohort, young people who at that time were roughly the same age as their parents were when they enrolled in the original Framingham study.

In 1976 a group of Harvard scientists, inspired in part by the success of Framingham, devised a similar long-term longitudinal study that looked at a much larger and much more highly select group of subjects: young female nurses. When it began under the direction of Frank Speizer of the Harvard School of Public Health and the Harvard Medical School, the study included a relatively small group of 30- to 55-year-old female nurses. The questionnaire used was designed to link their smoking history and use of oral contraceptives with subsequent disease. After the initial pilot study, the group was expanded to include more than 120,000 female nurses who every two years fill out a detailed six-page questionnaire about their health status, medical history, and habits regarding eating, exercising, smoking, drinking, and use of aspirin, birth control pills, replacement hormones, and vitamin supplements.

A subset of the participants in the Nurses' Health Study, about 32,000, submitted blood samples to be frozen and examined at a later date in light of medical problems they might develop. Another subset (68,000) submitted toenail clippings. The toenails, kept in a file drawer outside Walter Willett's office, were used to measure levels of the trace element selenium, which had once been thought—until the Nurses' Health Study showed otherwise[15]—to affect the risk of breast cancer.

Analyzing the data assembled from these tens of thousands of women requires sophisticated biostatistical maneuvers—not to mention time—to transfer the raw data into computers and then detect associations between, say, blood pressure, dietary fat, and alcohol consumption and the risk of colon cancer. From the extraordinary data base provided by the Nurses' Health Study, Harvard scientists have managed to turn much of conventional medical wisdom on its head. Their study subjects showed them that the risk of breast cancer is doubled among women whose mothers had premenopausal breast cancer;[16] that a high-fat diet significantly increases a woman's risk of colon cancer but not breast cancer;[17] that women who eat lots of fruits and vegetables rich in carotenes have a 40 percent lower risk of stroke and 22 percent lower risk of heart attack than women who do not;[18] that women who have one or two alcoholic drinks a day, in comparison to those who abstain, have a 30 percent lower risk of developing coronary heart disease;[19] that postmenopausal women on estrogen replacement therapy reduce their risk of coronary heart disease to premenopausal levels;[20] that women who gained more than 22 pounds since adolescence had significantly higher mortality rates in midlife;[21] and that women with diets high in *trans* fatty acids (partially hydrogenated vegetable fats, like margarine) have nearly twice the risk of heart disease as those with very low *trans* fatty acid intake.[22]

In addition to the obvious questions—such as "How many months has it been since your most recent mammogram?"—investigators with the Nurses' Health Study have asked some rather intimate (and less clearly health-related) questions, such as "How many close friends do you have?" and "How often do you attend religious services?" Anything about lifestyle, from exercise habits to social connections, has been considered fair game. At first, investigators focused on some of the women's health issues that were most prominently on people's minds in 1976: the health consequences for women of cigarette smoking, birth control pills, and hair dyes. Some of the results, especially those regarding smoking, were quite dramatic. The nurses who smoked were several times more likely to die of heart attacks than were the nonsmokers. Even a little bit of smoking did a lot of damage. Smoking just one to four cigarettes a day—a "light smoker" by anyone's definition—more than doubled a woman's risk of heart disease.[23] Heavy smokers (more than 25 cigarettes a day) had five and a half times the risk of fatal heart attack.[24]

For birth control pills the findings were less clearcut. In the mid-1980s investigators reported that the nurses who took birth control pills were no more likely to develop breast cancer than were their age mates who did not. But these findings came at a time when many other large-scale studies, such as the Cancer and Steroid Hormone Study of the National Cancer Institute, were finding an increased risk of premenopausal breast cancer among women who had ever used the Pill, especially if they had been on it for more than four years.[25]

Regarding hair colorings, the Nurses' Health Study found no link between hair dyes and various cancers. Even though hair dyes receded from the public's fear meter by the end of the 1970s, investigators kept assessing them for potential carcinogenicity. As late as 1994 they reported that nurses who dyed their hair did not have higher rates of lymphoma, leukemia, or multiple myeloma.[26]

As the times changed, the questions posed by the Nurses' Health Study also changed. In the 1980s, for instance, one important question related to the use of aspirin to prevent heart attacks. For years the Holy Grail of preventive medicine had been an easy way for people to reduce their risk of heart disease without sacrificing the foods they loved or embracing the exercise they dreaded. It seemed the grail was found in 1988, when aspirin was found to protect men from heart attacks. This finding came out of the Physicians' Health Study, a male-only follow-up study of 22,000-plus physicians who at the time were in their 40s through 80s. Charles Hennekens, a member of the Nurses' Health Study team, headed the Physicians' Health Study. After four years he and his colleagues found that the men who took regular-strength aspirin every

other day were nearly 50 percent less likely than the control group taking a placebo to suffer a heart attack.[27]

After the Physicians' Health Study results were publicized, the next
question became whether aspirin was equally beneficial for women. At
this point, the Nurses' Health Study began including questions about
aspirin use. By 1991 the Harvard scientists—who by now included Graham Colditz, Charles Hennekens, David Hunter, JoAnn Manson, Meir
Stampfer, Walter Willett, and more than 30 non-M.D. associates—were
able to report a significantly lower risk of heart attacks for women who
said they took aspirin on average one to six times every week. There was
no additional benefit to taking more aspirin and no benefit in terms of
stroke or overall cardiovascular mortality.[28]

In the late 1980s several large-scale studies suggested that a modest
weight gain throughout life was natural and inconsequential. Some
investigators, such as scientists working with the Baltimore Longitudinal Study sponsored by the National Institute on Aging, reported that a
slight weight gain was actually beneficial in terms of overall life expectancy. But by 1995 the Nurses' Health Study had uncovered evidence to
the contrary. Even a slight weight gain, on the order of 11 pounds at
any point in adulthood, was bad for women. The Harvard scientists
found that women who gained more than 11 pounds after age 18, no
matter how lean they were in adolescence, were 25 percent more likely
to develop heart disease than those who kept a consistent weight after
18—even if they started out at a somewhat heavier weight. Those who
gained more than 25 pounds doubled their heart attack risk. "Women
who were of average weight or somewhat leaner at age 18 and managed
to maintain their weight were the healthiest in the long run," says
Willett. Women who weighed just 10 percent more than the leanest
member of the study had a 20 to 30 percent higher risk of heart attack,
and those who were 25 to 45 percent heavier than the leanest women
had an increased risk of more than 100 percent.[29]

As the original cohort of nurses passed out of their childbearing
years—today they range in age from about 50 to about 75—questions
regarding the health hazards of hormones changed from looking at
oral contraceptives to evaluating postmenopausal estrogen replacement
therapy. Investigators are using data from the Nurses' Health Study to
determine the risks and benefits of estrogen replacement in relation to
heart disease (the number one killer of postmenopausal women),
osteoporotic fracture, and cancer (especially estrogen-related cancers
like breast and uterine cancer). Over the years a picture has emerged
that estrogen replacement is, on balance, beneficial for most women.
The postmenopausal nurses on estrogen have half the heart disease risk
of their age mates who do not take estrogen, as well as a significantly
reduced risk of osteoporosis and no increase in uterine cancer. The
primary negative has been a 30 to 40 percent increase in breast cancer
risk among the nurses on estrogen replacement therapy.

So how does an individual woman evaluate the risks and benefits of estrogen? How does society at large? "My view is that for the average woman, the benefits outweigh the risks," says Meir Stampfer of the Nurses' Health Study team. "But there are plenty of women for whom the reverse is true. Ideally, before deciding whether to take hormones, a woman's risk profile should be identified for all the diseases related to estrogen: heart disease, breast cancer, and osteoporosis."[30]

From a public health perspective, though, the issues to be considered go beyond a woman's risk profile. On a population level, what matters is whether more people can be expected to encounter a particular therapy's positive effect than its negative effect. One way to make such an assessment regarding hormone replacement therapy is to look at a hypothetical group of 2,000 postmenopausal women in any given year. Of these 2,000, if no one takes estrogen, 20 will develop heart disease and 12 will die of it; 11 will break an osteoporotic bone and one or two will die of it; six will develop breast cancer and two will die of it; and three will develop uterine cancer and one will die of it. If all 2,000 women were on estrogen replacement therapy, seven of those destined to die of heart disease or osteoporosis would be spared, while one additional person would die of cancer who wouldn't have otherwise. Seven lives saved versus one life lost makes estrogen replacement therapy seem a good bet from the perspective of an entire population.[31]

Today the Nurses' Health Study is a major research project at Harvard, based at the Brigham & Women's Hospital in Boston. Running on about $2 million a year in federal grants, involving nearly 40 investigators, generating more than 120 research papers since its inception, the Nurses' Health Study continues to collect and analyze reams and reams of data. At the program's offices, as many as 10,000 completed questionnaires at a time arrive in the mail.

Like the Framingham Heart Study, the Nurses' Health Study has spawned offspring of its own. Besides the original 120,000 nurses, a new cohort of 116,000 younger nurses also is being followed. In addition, Harvard researchers have recently broadened their base to include men. These days, another 50,000 male health care professionals (dentists, veterinarians, optometrists, and pharmacists) are also filling out lengthy questionnaires.

Population studies like Framingham and the Nurses' Health Study have all pointed to a set of risk factors that contribute to heart disease and other causes of morbidity and mortality. Some are innate factors with big payoffs that are impossible to change: one's family history, for instance, or one's sex, or the fact that the risk of heart disease almost always rises with age. Some of these are behavioral factors that can be changed—certain dietary habits or one's level of physical activity—though the effects of making any single change tend to be relatively small. But one important risk factor is very much within people's control, and eliminating it can have an enormous impact on the likelihood

Providing Guidance for Individual Behavior

of developing many of the illnesses plaguing the industrialized world. That risk factor is, of course, cigarette smoking. Because it has such profound effects on the public's health, officials have focused their attention on it quite steadily for the past 30 years, ever since the dramatic connection was originally made between smoking and disease.

Smoking: The Deadliest Habit

Until the mid-1950s, nobody worried much about smoking. In part, this was because tobacco in its most potent form—cigarettes—was relatively rare until the invention of the automatic cigarette roller. Early in the twentieth century, Americans smoked an average of only 1,000 (hand-rolled) cigarettes per year per person—which translates in modern parlance into a little less than a pack a week. (In contrast, the average smoker in 1977 went through 12,854 cigarettes a year—more than a pack and a half each and every day.[32]) And tobacco as it was usually consumed, in pipes and as snuff, was thought to be mostly harmless.

Even in the middle of the twentieth century, the general consensus was that cigarettes were not that bad for you. When the original advisers to the Framingham Heart Study planned their questionnaires in 1948, most were genuinely unsure whether smoking would be shown to have any effect on the risk of heart disease. As Tom Dawber puts it:

> Most of us weren't at all sure about the effects of cigarette smoking. We looked at it only because we thought we ought to look at all the environmental factors we could, especially since we were not really in a position to do genetic studies anyway.[33]

This was, after all, just a few years after the medical profession had approved the inclusion of cigarettes in the ration kits sent to young men on the front lines of World War II. Physicians endorsed the concept of supplying tobacco to soldiers, because they considered it a highly effective way to relieve tension. Who knows how many young men became nicotine dependent only because government-issue cigarettes were included in their rations?[34]

By the 1950s, however, the professional attitude toward smoking began to change. The main impetus was a 1952 study conducted in Britain by two London epidemiologists, Richard Doll and Bradford Hill. They reported a dramatic correlation between cigarette smoking and lung cancer. From then on, recalls Brian MacMahon of the Harvard School of Public Health, the relationship between smoking and lung cancer was "clear as day."[35]

After the Doll-Hill study was reported, similar studies from around the world confirmed this dramatic association. Some public health officials in this country then began making pronouncements about the health risks of cigarettes. In 1956, U.S. Surgeon General Leroy Burney published an article in the *Journal of the American Medical Association* concluding that "the weight of evidence at present implicates smoking as the principal etiological factor in the increased incidence of lung cancer."[36] This article received little attention, though, possibly because neither Burney nor anyone else in the Eisenhower White House took up the antismoking banner as a personal cause. But Burney's successor did—as has every surgeon general since.

Luther Terry was the first surgeon general to attack cigarettes with a vengeance. Terry, a cardiologist who had previously been deputy director of the National Heart Institute, convened the 10-person Advisory Committee on Smoking and Health, which met regularly between November 1962 and December 1963. Even the assassination of President Kennedy did not derail these scientists from addressing their subject with vigor. During its 13-month tenure, the committee reviewed all available animal, clinical, and epidemiological evidence on smoking and health. When looked at from this comprehensive perspective, the news was grim indeed. Terry presented the panel's findings in January 1964 at a press conference he deliberately scheduled for a Saturday to minimize its impact on tobacco stocks. Before a standing-room-only crowd of reporters, the surgeon general announced:

> Cigarette smoking is causally related to lung cancer in men. The magnitude of the effect of cigarette smoking far outweighs all other factors. The data for women, though less extensive, point in the same direction.[37]

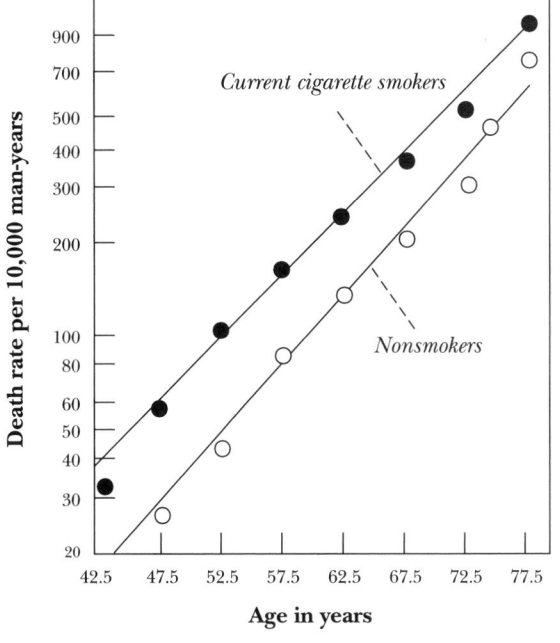

Death rate (logarithmic scale) plotted against age. Prospective study of mortality in U.S. veterans. From the landmark Surgeon General's report, Smoking and Health, *1964.*

From then on the Office of the Surgeon General became synonymous with antismoking. For Terry and his successors it sometimes must have seemed that every interaction with Congress, the press, and the medical community related in some way to cigarettes. "When I testified in Congress there were major attacks on me," recalls William Stewart, surgeon general from 1965 to 1969. "The man from North Carolina came into the hearing with a tobacco plant in a pot and he said, 'This man is trying to kill this beautiful plant.' "[38]

It was under Stewart's direction that the warning on cigarette packages, mandated by the Federal Trade Commission, was strengthened. Stewart approved text that was direct and unequivocal: "Warning: The Surgeon General Has Determined That Cigarette Smoking Is Dangerous To Your Health." More than 20 years later, though, even prose that had once seemed dramatic had become so commonplace that smokers generally ignored it each time they ripped open a new pack of cigarettes. That is when a new surgeon general, C. Everett Koop, changed tactics. Koop's staff designed four different labels that specified the effects of tobacco on either the heart, the lungs, the fetus, or one's overall health. The four alternative warning labels rotated from one cigarette pack to another, so that smokers might actually notice, and maybe even read, words that had not yet become invisible because of familiarity.

Since Terry's first report, the federal government has tried several approaches to wean Americans from cigarettes. In 1966 the Federal Trade Commission ruled that tobacco advertising must not be aimed at people under age 25. The following year the Federal Communications Commission imposed an "equal time" regulation: for every minute of cigarette advertisements on radio and television, the station must also broadcast one minute of antismoking messages. Finally, in 1971 Congress banned any cigarette advertising on electronic media.

This last action might have had a paradoxical effect. Because the outright ban also did away with the useful public health messages that had been delivered under the equal time provision, many experts believe the ban ultimately backfired. As Carl E. Bartecchi, Thomas D. MacKenzie, and Robert W. Schrier of the University of Colorado School of Medicine recently wrote:

> During the three years following the electronic media ban, per capita cigarette consumption actually rose slightly before resuming its drop. Many analysts attribute the brief surge to the cessation of public service announcements that coincided with the electronic ban.[39]

As tobacco advertisements shifted to print media, especially magazines, the ban backfired in a different way. Not only were the magazine advertisements unanswered by any kind of equal-time public service announcements in the print media, but the presence of the advertisements actually poisoned the editorial content of the magazines in which they appeared. In the first three years after the broadcast ban, according to Bartecchi, MacKenzie, and Schrier, the magazines that carried the most cigarette advertisements reduced their coverage of smoking-related health issues by 65 percent, compared to a decline of 29 percent in magazines that did not carry cigarette ads.[40] And despite the electronic ban, tobacco companies ultimately found their way back onto television, primarily by sponsoring sporting events.[41]

In the meantime, study after study documented the adverse health consequences of cigarette smoking. From a habit that people at first linked conclusively only to respiratory diseases—primarily bronchitis, emphysema, and lung cancer—epidemiologists were showing that smoking increased the risk of many other cancers (especially bladder, pancreatic, and esophageal), as well as heart attacks, strokes, hypertension, osteoporosis, and peptic ulcers. Yet, maddeningly, cigarette consumption continued to rise, especially when measured as the number of cigarettes smoked by the average smoker. In 1977 this statistic peaked at 12,854 cigarettes per smoker per year. This was largely an artifact of the dropping from the smoking pool of those who smoked the least and who therefore found it easiest to quit once they heard how dangerous cigarettes were. But this increase in average cigarette consumption made it clear that a new strategy was needed, since 13 years' worth of antismoking messages by the surgeon general, the U.S. Public Health Service, professional medical associations, the National Institutes of Health, schools of medicine and public health, and organized public interest organizations apparently had little effect on the people who were most seriously hooked.

It was time, the experts concurred, to change the way that smoking itself was perceived. In the 1950s and 1960s smokers were thought to be sophisticated and elegant; movie stars were photographed with cigarettes dangling from their lips or perched between their fingers. But by the 1970s smoking was on its way to seeming socially obnoxious. Chain-smoking movie stars like Humphrey Bogart were dead of lung cancer, information about the health hazards of smoking was becoming well known, and in some circles smokers no longer looked sophisticated; they just seemed ignorant, short sighted, and hopelessly addicted.

By the 1980s the national mood toward smoking was ready for a dramatic shift. This occurred when experts learned more about the dangers of second-hand smoke. The first demonstration of the health hazards of passive smoking was a case-control study conducted by epidemiologists at the Harvard School of Public Health and published in 1981, led by former chairman Brian MacMahon and his student and successor as chairman, Dimitrios Trichopoulos. Trichopoulos found 51 women who were hospitalized with lung cancer in his native Greece and compared them with 163 age-matched controls who were hospitalized at the same time for other reasons. None of the women in the study smoked. Trichopoulos and MacMahon found that the lung cancer patients were significantly more likely than the others to have been exposed to smoke through their husband's cigarettes.[42]

This was the first in a series of studies that concluded that a cigarette smoker's toxic presence constitutes a health hazard to children and other nonsmokers who are nearby. With these findings, smoking was no longer perceived as merely a bad habit that endangered an individual's own health status. It became instead a public health menace that endangered everyone.

The mobilization of nonsmokers gained momentum in 1989, when Surgeon General Koop put a twist on the annual *Surgeon General's Report on Smoking* by focusing his analysis on the emerging concept of passive smoke. Koop's report revitalized the antismoking

Dimitrios Trichopoulos

effort and gave activists important new ammunition to carry their campaign into the 1990s. No longer could smokers say "It's my body. I can do what I want to it. I'm not hurting anyone." Koop's report made such a line of thinking untenable by stating that, for every hour a nonsmoker spent in a smoke-filled room, he or she encountered the same lung damage as would have occurred from smoking a cigarette.[43]

In the years since the surgeon general's report on passive cigarette smoke, new information has poured in about the health hazards of involuntary exposure to tobacco. The Environmental Protection Agency, for instance, estimated that side-stream smoke—the smoke that moves from a lit cigarette into the surrounding air—causes 3,800 lung cancer deaths a year, making it as potent as a class A carcinogen.[44] Armed with such statistics, activists have pushed through a series of antismoking bills in the 1990s. The phrase "smoke-free environment" has spread to apply to all domestic airline flights, all federal buildings, and many offices and restaurants nationwide.

But just as the antismoking lobby has been vigorous in its efforts, so has the tobacco lobby—and with far more resources at its disposal. The industry donated more than $4.7 million during the 1992 presidential election to all the leading candidates, tripling its contribution from 1988. And the long arm of the lobbyists extended to most congresspersons as well. In the late 1980s, according to one calculation, 420 of 535 members of the House of Representatives, and 87 of 100 senators, accepted campaign contributions from the tobacco industry.[45]

The aggressive methods of the tobacco lobby have been an eye opener for many idealistic public health advocates. William Foege, for instance, says he cannot understand the motivation of people who want to push a dangerous substance, knowing full well the illnesses it can cause. Foege—who was trained at the Harvard School of Public Health in the mid-1960s—had his earliest public health experiences battling smallpox in the early 1970s, when the enemy was clear and unambiguous. But when he served as director of the Centers for Disease Control under Presidents Carter and Reagan from 1977 to 1983, he came up against the tobacco lobby, which he recalls was unlike any public health foe he had ever confronted.

> When I got into public health, you were never fighting people. There wasn't a tuberculosis lobby or a malaria lobby. There weren't people making money on seeing diseases spread. So it was a very sobering thing to get into the whole tobacco issue and find out that there are people who, even faced with the evidence and the facts, knowing that their decision to advertise and promote tobacco is going to kill people, knowing all of that, are still willing to do it. One sometimes yearns for the old public health days, where your enemies were honest.[46]

The general trend in smoking rates has been downward in the past generation, but progress has been uneven. The statistics on deaths and disease from tobacco remain sobering. In 1990, 19 percent of all deaths in the United States could be traced to tobacco-related illnesses. This proportion climbed to 25 percent for deaths among people ages 35 to 64. An estimated 400,000 Americans die each year from smoking; when deaths from all causes attributable to passive smoke are included, the figure rises to almost half a million.[47]

Fewer women than men have stopped smoking, and young people—an important indicator of the persistence of the habit into the next century—have been the most intractable group. According to Antonia C. Novello, surgeon general under President George Bush, the percentage of male smokers dropped by nearly 30 percent between the mid-1950s and the early 1990s, but the percentage of female smokers fell by only 4 percent. And while the overall proportion of adult smokers has been falling, the proportion of teenagers who smoke has held steady: 18 percent for teenagers of all races and a shocking 22 percent for white teenagers.[48]

To have a real impact on smoking rates, the most valuable step is to stop young people from becoming smokers in the first place. The best way to do that is to increase the price of a pack of cigarettes. Studies have shown that teenagers are particularly sensitive to cost: for every 10 percent increase in cigarette prices, there is a reduction in smoking rates of about 4 percent, most of it among teenagers. An increase in prices would have a long-term impact on overall rates of smoking, then, because some 90 percent of adult smokers got hooked on cigarettes before the age of 21.[49]

Public health statisticians tend to think that if people could only see, in cold hard numbers, how dangerous their habits are, they might be convinced to change them. So they make calculations like the one that states that every cigarette shortens a smoker's life span by five and a half minutes—about the time it takes to smoke it.[50] Another statistic: if tobacco were eliminated altogether, every current smoker would gain on average an extra four or five years of life. For smokers destined to develop a fatal tobacco-related illness—admittedly, a subset that is impossible to identify in advance—each of them would gain an astounding 15 years of life if they lived in a society in which there were no such thing as cigarettes.[51]

In the past 20 years, public health officials have tried to issue similar pronouncements about a bad health habit even more pervasive in the United States than smoking: poor nutrition, which in America today does not mean malnutrition so much as overnutrition. Although no single factor in an individual's diet contributes to ill health as significantly as do cigarettes, eating wisely remains the next most important thing an individual can do to ensure a long and healthy life.

Making Healthy Choices Three Times a Day

As the Nurses' Health Study evolved over time in response to the major health questions of the day, it shifted imperceptibly to a greater and greater focus on nutrition. At the instigation of team epidemiologist Walter Willett, who subsequently became chairman of the Department of Nutrition at the Harvard School of Public Health, questions about eating habits were first included in the nurses' questionnaire in 1980. Today, in the biennial survey, investigators ask how often respondents eat particular foods—daily, weekly, once or twice a month, never—and pose such questions as, "Do you currently take multivitamins?" "How many teaspoons of sugar do you add to your beverages or food each day?" "What kind of cold breakfast cereal do you usually eat? Specify brand and type. What kind of fat do you usually use for baking at home?"[52]

Of the different ways to assess the relationship between diet and disease, the Nurses' Health Study represents the prospective-cohort approach. This strategy, in which a large sample of individuals are asked about their eating habits before and as they get sick, has fewer complications than do alternative methods of studying large populations. The case-control study, which compares the eating habits of a group of people who are already sick with the habits of a comparable group of people who are healthy, has several potential drawbacks. The control group might be different in some undefinable way from the case group.

Moreover, people's recollections of exactly what they ate in years past might be biased, especially among the case subjects, whose memories might be tainted by the fact that they eventually developed the disease under study. Another type of study, the correlational study, compares diets and disease rates in several different countries. Its weakness is that whatever differences are observed in disease rates from one country to another might be explained by some lifestyle, genetic, or environmental factor quite unrelated to diet.[53]

The Nurses' Health Study, one of the largest and oldest prospective cohort studies in the world, avoids the pitfalls of many other study designs. This explains why its nutritional findings carry such weight. Among the dietary conclusions are many that went against the conventional wisdom of the day when they were reported:

- *Dietary fat may not be related to breast cancer.* A low-fat diet in adulthood affords no protection against the development of breast cancer, and a high-fat diet does not seem to promote breast cancer.[54]

- *Red meat may not hurt a woman's heart.* Women who eat red meat do not have a higher risk of heart disease, possibly because the harmful component of red meat for men is its high iron content, which is not a problem for premenopausal women, who are often iron deprived.

- *Red meat* does *hurt a woman's colon.* Women who eat red meat (beef, pork, or lamb) once a day have more than twice the risk of colon cancer than those who eat it only once a month.[55]

- *A little bit of alcohol may be beneficial.* Women categorized as moderate drinkers (three to nine drinks a week) have half as much coronary heart disease as teetotalers—a finding that even the researchers called "remarkable."[56] However, the same level of alcohol consumption is also associated with a modest increase in the risk of breast cancer.

- *Even a little produce is good for you.* Women who eat as little as one serving of fruits or vegetables every day have a decreased risk of heart disease and stroke compared to women who eat less.[57]
- *Margarine may be as bad for you as butter.* Women who consume the highest amounts of *trans* fatty acids (found in hydrogenated vegetable oils, primarily margarine) are nearly twice as likely to have a heart attack as women who consume the lowest amounts.[58]

The margarine finding was released in mid-1994 to a chorus of editorializing along the lines of "Well, then, what *is* safe to eat?" It is a graphic demonstration of how difficult it is to make intelligent dietary recommendations at a point in time when available knowledge is incomplete. "If the experts were going to flip-flop on margarine—sensible, virtuous margarine," wrote journalist Katherine Griffin in a typical opinion column, "why should we believe them about anything?"[59]

The margarine finding might have caused consternation in the general public, but that was nothing compared to how distressing it was for the people compelled to issue it. Many of them, after all, were the very people who advised Americans to reduce their consumption of butter, a recommendation that in turn led to a sharp rise in the use of margarine as a substitute. "A major artificial element has been introduced into the food supply without a full understanding of all its metabolic and health implications," wrote Willett and his associate, Alberto Ascherio, in an editorial in the *American Journal of Public Health*. Estimating that as many as 30,000 Americans a year may be dying as a result of eating too many *trans* fatty acids, the Harvard scientists recommended that people cut way back on their consumption of *trans* fatty acids. Margarines without *trans* fatty acids, commonly available in Europe, may be part of the answer. As Willett and Ascherio wrote:

> Prudence suggests that we adopt a low threshold for evidence of harm for synthetic substances added to the food supply that have no known nutritional benefits. We believe that the threshold of evidence for harm has been far surpassed in this case; the metabolic data alone should be a sufficient basis for limiting human intake of partially hydrogenated vegetable fat.[60]

The problem of determining the precise contribution of diet to disease has haunted nutritional scientists for centuries. Take scurvy, one of the first vitamin deficiency diseases to come to anyone's attention. As long ago as the seventeenth century, medical authorities recognized that the mysterious illnesses and deaths among crew members on long sea voyages—men who were perfectly healthy when they set sail—might be related to a lack of citrus fruits on board. Indeed, the first time that lemons and oranges were recommended to cure scurvy was in 1611.[61]

Other cures were recommended, too, from sunlight to sauerkraut. Not until 1753, when legendary British naval surgeon James Lind conducted a controlled experiment on members of the Queen's navy, did anyone prove that fresh citrus fruits really do prevent scurvy. At first, naval authorities resisted the finding; they were hoping not to have to take the expensive step of providing a daily ration of fresh citrus fruit to all sailors on extended voyages. In the end, the admiralty's stinginess almost undermined the entire scurvy prevention effort. Naval procurers found citrus at a good price, in the form of limes from the West Indies. But the cut-rate limes were low in vitamin C, and even though they gave the British their enduring nickname of "limeys," they did little to prevent disease. Sailors continued to develop scurvy on long sea voyages as late as 1875—more than a century after James Lind's definitive finding.[62]

Nutrition research has grown steadily more sophisticated in this century, although much still depends on controlled clinical trials, careful observation, and the time-consuming work of food diaries and dietary manipulation. The high-technology aspects of nutritional research involve biochemical analysis of foods, as well as work with experimental animals. This latter kind of research has shown, for instance, the connection between dietary cholesterol and high cholesterol in the bloodstream, known as hypercholesterolemia.

In the 1940s no one considered using animals to investigate the relationship between diet and hypercholesterolemia, primarily because no one thought laboratory primates developed atherosclerosis (narrowing of the blood vessels) in a way comparable to humans. But then no one had really tried to induce it, says Fredrick Stare, who founded the Department of Nutrition at the Harvard School of Public Health in 1942 and chaired it until his retirement in 1976. Stare and his colleagues were the first scientists to create an animal model for hypercholesterolemia (high cholesterol in the blood) in several species of New World monkeys. These monkeys, particularly cebus and squirrel monkeys, proved to be enormously helpful in explaining how high cholesterol contributes to atherosclerosis.

Fredrick Stare, founding chair of the Department of Nutrition, 1946

One of the puzzling aspects of atherosclerosis had been the occasional discrepancy between hypercholesterolemia and the incidence of atherosclerosis or ischemic heart disease [caused by lack of blood flow to the heart]. Whereas hypercholesterolemia is considered by many to be a primary risk factor in atherosclerosis, some individuals with elevated levels of cholesterol do not develop significant atherosclerotic vascular disease while other normocholesterolemic persons do.[63]

Stare's findings helped clear up this apparent paradox. A diet high in saturated fat, in the form of coconut oil, had different effects on different species of monkeys. Both the squirrel and cebus monkeys experienced a rise in blood cholesterol, but the squirrel monkey, which seemed to develop less heart disease than did the cebus monkey, experienced most of its increase in the form of high-density lipoproteins (HDL). As Stare wrote later, these findings

> support the concept of the protective nature of high-density lipoproteins and the atherogenic potential of low-density lipoproteins (LDL). They also suggest that a species genetic control of the lipoprotein response to diet is variable and has important biologic implications.[64]

In midcentury most public health scholars, including Stare and his associates, felt that research into the basic science of nutrition did not go far enough. After all, their mission was to improve the health of entire populations and that meant an equal emphasis on nutrition education. That is why one of Stare's proudest accomplishments was the creation of a catch phrase that has become second nature to most Americans raised in the 1950s and 1960s: the Basic 4 Food Groups.

> The Department of Agriculture had the Basic 7, but we wanted to make it simpler. I think in dealing with students, the simpler you can keep it the better. We proposed something that we called the Basic 4, which gave us a nice graphic of a shield so we could say "Protect your health with protective foods." We combined the fruits and vegetables groups—for all practical purposes, the nutrients that you get out of fruits and vegetables are the same, so why have two separate groups?— and we also combined anything made out of cereal into one group: it can be bread or noodles or pasta or potatoes or cornflakes, whatever you want. Then we had high-protein foods—meat, poultry, eggs, legumes, fish—and, last, anything made out of milk.[65]

Stare's basic four food groups were revised recently by the U.S. Department of Agriculture (USDA) and by some academic scientists, including Stare's own successors at the Harvard School of Public Health. The USDA favors a new graphic known as the food pyramid, which recommends many portions a day of the food groups at the pyramid's base (grains and complex carbohydrates) and a few portions a week of the foods at its apex (sweets and other sources of empty calories). Walter Willett, the very man who holds the Fredrick John Stare Professorship of Epidemiology and Nutrition at the Harvard School of Public Health,

Stare's Basic 4 Food Groups

is one of the most vocal critics of the USDA's food pyramid. Willett, who earned a doctorate in public health at Harvard in 1980, favors an alternative dietary balance represented by an alternative pyramid: the Mediterranean diet food pyramid.

Willett's pyramid differs from the USDA's in several important ways. Like the USDA version, Willett's model emphasizes the foods at the pyramid's base (fruits, vegetables, and whole grain breads, pasta, rice, couscous, polenta, bulgur, and potatoes) and deemphasizes the foods at the top. But the Mediterranean diet rearranges which foods are at the top (eggs, sweets, and, at the very apex, red meat), and has a separate category for olive oil.

For his part Stare defends the value of the simplest, oldest advice, the Basic 4 and its handy shield.

> The Basic 4 has been around since we first published our paper on it in 1955. The USDA came out with its own Basic 4 a year later, and even though it was the same as ours, we usually don't get any credit for it. But that Basic 4 is now in most all biology texts in college and in high school, so why change it? Why confuse people? The USDA says we need more than four food groups, that now there should be five or six. But I say keep it simple.[66]

As evidence continues to accumulate from large population studies regarding which foods are healthiest and which should be avoided, keeping it simple becomes an ever-more elusive goal. A similarly confusing pendulum swing has occurred in the prevailing professional opinion about one component of many Americans' diets—alcohol.

The prosperous 1950s, especially in the burgeoning suburbs of America, were often characterized as one endless round of martinis. In his famous short story, "The Sorrows of Gin," John Cheever describes a typical Friday evening in the Connecticut town of Shady Hill: the protagonist, Mr. Lawton, drinks one martini before climbing aboard the commuter train, makes a shaker of martinis to share with his wife as soon as he walks in the door, has another before heading out to dinner with the neighbors, has several more drinks while he is out, and then, when he returns, he has one more for the road before taking the babysitter home.[67]

It is perhaps no coincidence that the 1950s were also the decade when the public health and medical communities were waking up to the stunning health impacts of alcoholism. In 1951 the American Public Health Association presented a Lasker Award to Alcoholics Anonymous "in recognition of its unique and highly successful approach to that age-old public health and social problem, alcoholism." The same year the American Medical Association created a special subcommittee of its Committee on Chronic Diseases to make a plan for solving the problems of alcoholism through medical intervention. This was also the time when public health officers first defined alcoholism as a disease:

> a chronic and progressive condition characterized by an ever-growing dependency on alcohol with loss of self-control as to quantity and occasion. It is accompanied by manifestations of physical malfunction and psycho-social maladjustment, particularly in later stages.[68]

In the decades since, experts have been quick to recognize that alcoholism accounts for over 100,000 excess deaths per year. Alcohol-related deaths are the third or fourth leading causes of death for all races and both sexes (in between strokes and traumatic injury).[69] Most often alcoholics die because their livers give out, but sometimes they suffer from brain damage or from cancers of the mouth, throat, esophagus, or rectum. Besides that, the compulsion to drink each and every day takes over the life of the alcoholic. As one recovering alcoholic put it:

> I did not choose to be an alcoholic. I did not choose the blackouts, the job disasters, the collapse of my carefully constructed life. . . . The beginning of recovery was mysterious. On a Sunday morning, after a long night of drinking, I was cold sober. I sat at a kitchen table in Staten Island, asking for another drink, and suddenly I felt my life to be unbearable.[70]

One of the most pernicious forms of alcoholism is compulsive binge drinking. Binging—defined as five drinks in a row for males, four for females—has become so pervasive on college campuses that many observers consider it to be an emerging public health catastrophe. "We can no longer dismiss binge drinking as young people's games," says Henry Wechsler of the Harvard School of Public Health, "because a significant number of students get involved in serious problems."

Wechsler, who heads the College Alcohol Studies Project at Harvard, surveyed 17,500 students at 140 colleges and universities in 1993 and found that almost half were binge drinkers. Binge drinkers are five times more likely to encounter troubles associated with alcohol—unplanned sex, blackouts, academic problems, trouble with the police—than their more moderate classmates. The average binge drinker is male, white, athletic, and a member of a fraternity; he typically has lots of friends, a roommate, and a parent who graduated from college. The average nonbinger is religious, serious about studies, involved in the arts or community service, and employed part time.[71]

While public health experts agree on the detrimental effects of excessive drinking and alcoholism, they are less sure about the effects of light to moderate drinking. In fact, many studies suggest that small amounts of alcohol actually protect a person's health. In late 1994 two public health experts from New York concluded that the deaths averted through moderate alcohol use might nearly balance out those caused by alcohol abuse. In an editorial in the *Journal of the American Medical Association,* Thomas A. Pearson of the Bassett Research Institute in New York and Paul Terry of the Columbia University School of Public Health wrote:

> The key is to tailor the message [regarding alcohol consumption] to each individual, in the same way that counseling is given on diet, physical activity, sexual practices, and so on. Certainly, there is a subgroup of the population who should not consume alcohol at all. Identification of these persons is especially important in the second and third decades of life, when the excess of alcohol-related deaths occurs and when alcohol drinking patterns become established.[72]

Most of the studies conducted in recent years have concluded that, at least for men, a light-to-moderate intake of alcohol reduces overall death rates, largely because it reduces the death rates from coronary heart disease. But the conclusions for women have been more complicated. While light to moderate drinking probably has a similar beneficial effect on women's rates of heart disease deaths, several studies have found that it leads to a slight increase in deaths from breast cancer—an increase that might offset any potential benefit.

In 1987, for instance, researchers with the Nurses' Health Study found a clear association between alcohol consumption and breast cancer. They concluded that daily drinkers had a 40 percent higher risk of breast cancer than did nondrinkers.[73] A similar association was reported the same year by the National Cancer Institute in its National Health and Nutrition Examination Survey.[74]

More recently, though, an analysis of the Nurses' Health Study examined alcohol intake in relation to total mortality and specific causes of death. Yes, alcohol does increase the risk of breast cancer (and of cirrhosis, too, to which female drinkers seem particularly susceptible), but this greater risk is usually seen in women who consumed two to

155

three drinks every day. For lighter drinkers the breast cancer death risk was not significantly increased, while the risk of death from coronary heart disease was significantly reduced. As the Harvard researchers concluded, "the apparent benefit of light-to-moderate alcohol consumption was mainly confined to women at greater risk for coronary heart disease, specifically older women and women with one or more coronary risk factors."[75]

For all the debate over the pros and cons of light drinking, there is still little question that one of the biggest problems with alcohol is its role in car crashes and other sources of injury. To change people's behavior as it relates to drunken driving, though, takes more than a traditional public health message stating that alcohol is bad for you. What is required and what has been tried on a large scale in the past 10 years is a deliberate effort to change the entire social context in which drinking and driving occur—making drunken driving as shameful and socially unacceptable as stealing.

The "Designated Driver" and "Squash It!" Campaigns

In October 1985 Dennis Kauff, a popular Boston newsman, was killed by a drunk driver. At his funeral, attended by more than 500 people—including many leaders in the local news media—the anchorman on Kauff's TV news show observed in his eulogy that "there's so much anger in this room, it could blow the roof off the place." That remark, reported the next day in Boston newspapers, caught the attention of Jay Winsten of the Harvard School of Public Health.

At about this time, Winsten, director of the school's new Center for Health Communication, had grown enamored of an idea, which originated in Scandinavia, that fought drunk driving without attacking drinking itself. The idea was that a designated driver would be chosen at the

Felton Earls (left) and Jay Winsten (center) at conference on violence and mass communication

beginning of a party or social event with the understanding that that person would drink no alcohol at all and thus safely drive everyone else home. Winsten had first heard of the concept through an initiative sponsored by the Washington (D.C.) Regional Alcohol Program. Winsten liked it so much that he went to Sweden to see whether it could be exported to the United States on a national scale.

In Sweden, says Winsten, "there's a very powerful social stigma connected to an arrest for drunk driving." He recalls an epiphany in a popular Stockholm bar, the Dixie Queen, built in an authentic Mississippi riverboat moored in the Gota Canal, where he went "to conduct field research."

> I put down my 50 kroner and went on board and struck up some conversations at the bar. I met a group of four young men who had come together by taxi, 20 U.S. dollars each way—and this was typical; at closing time, the lineup of taxicabs stretched around the block. One of the men, a housepainter, said to me, "If I were to get drunk and then drive, my friends wouldn't speak to me, and my brother would beat me up." Now that's a social norm that goes far, far beyond the risk of arrest and conviction and incarceration for drunk driving.[76]

What Winsten especially liked about the Swedish notion of the designated driver was its simplicity. It was a positive message—rather than the typical negative phrase "Don't drink and drive"—that could be reduced to a few words. Perhaps more importantly, it promoted a social norm that the driver doesn't drink and gave social legitimacy to the person who chose not to drink. When a host offered an alcoholic beverage, it was a simple matter to say, "No thanks. I'm the designated driver."

At first, Winsten promoted the designated driver concept in the United States through restaurant promotions, encouraging proprietors to offer free soft drinks to designated drivers, and prime-time public service announcements, particularly during holiday seasons. But one of Winsten's advisers was telling him that he had to do more, that the concept had to be inserted into TV shows themselves. The adviser was Frank Stanton, former president of CBS.

> Dr. Stanton and I would meet periodically for breakfast at the New York City Harvard Club, and I would tell him what we were doing as far as promoting the designated driver idea, mostly through the print media and through public service announcements during prime time. And he would say, "You're doing great work, but you're missing the boat. Nothing can begin to rival the potential impact of entertainment programming."[77]

By inserting into an evening sitcom or drama just a few lines of dialogue that highlighted the idea of a designated driver, Stanton argued that the concept would be brought to a different and potentially huge market. "He pointed out that viewers closely identify with particular characters, so there's enormous potential for modeling behavior," Winsten recalls. "Now, I come from molecular biology and the lab, and I didn't take instantly to this notion of Hollywood. In fact, it took Dr. Stanton quite a bit of time to convince me to try it."

What ultimately developed was a project that has been described as the first collaboration between Harvard and Hollywood since "Love Story."[78] Stanton and his close friend Grant Tinker, former chairman of NBC who had just become an independent Hollywood producer, took the idea under their wing and made connections on Winsten's behalf. Tinker wrote personal letters of introduction for Winsten to the CEOs of the 13 largest production companies in Hollywood, and he had them RSVP directly to Tinker about whether they would meet with Winsten. If they didn't call him, Tinker would call them. All 13 eventually agreed to meetings in April 1988, and all agreed to participate in the project.

After that initial round of interviews, Winsten spent 25 weeks in Los Angeles meeting with 250 executive producers and chief writers from scores of television programs.

> Many of the producers were of an age where they had teenage kids of their own, so when you point out that drunk driving is the number one cause of death among young adults in the United States, you've won their attention. Many of them have stories to tell, if not about their kids then about friends of their kids.[79]

Almost all of the producers agreed to cooperate. Winsten thinks they were partly "intrigued" by the Harvard connection, but more than that they were convinced of the importance of what they were being asked to do, and what they were being asked to do was relatively easy to accomplish. As he says:

> I asked them to consider, on an occasional basis—and only if from their point of view it worked for their show—incorporating a line or two of dialogue to reflect a new social norm about drinking and driving, which I argued on the basis of data was already beginning to evolve. I wasn't asking them to portray a social norm that was the opposite of reality, but rather to reflect changes already under way and thereby add momentum to those changes.[80]

The results could be seen by the 1988–1989 TV season. On one episode of the drama series "Hunter," for instance, a waitress brings drinks to four detectives sitting together at a bar. "So who had the soda water?" she asks, and one of the macho guys says, "I did." "Well," the waitress smiles, "someone's gotta drive." Such simple exchanges occurred again and again on TV shows—so often, in fact, that people practically stopped noticing. As Winsten puts it:

> Through four television seasons, by our count—and then we got tired of counting—160 programs did as we asked. Either a line or two of

dialogue or an entire scene or a subplot woven throughout the show dealt with drunk driving or designated drivers, or, in over 30 instances, an entire 30- or 60-minute episode dealt primarily in its plot with drunk driving and its consequences. The trade press estimated that the combined economic value of the prime time public service announcements plus the time in the shows was in excess of $100 million a year.[81]

By the early 1990s, "designated driver" was so much a part of the American lexicon that it had become an entry in the *Random House Webster's College Dictionary* (1991 edition). The Roper Poll tracked the behavioral changes associated with this new familiarity. By 1989, according to the poll, 29 percent of adults over age 30, and 43 percent of adults under 30, had been designated drivers themselves; by 1993 those figures grew to 39 percent for the older group and an impressive 56 percent for the younger group. Ninety percent of all adults surveyed approved of the designated driver concept as a way to end drunk driving.[82]

A more important measure of success has been in the reduction of deaths from drunk drivers. Between 1988 and 1994, the annual number of alcohol-related traffic fatalities fell by almost 30 percent.[83] In addition to the designated driver campaign, this decline was partly due to changes in the legal drinking age, which by 1988 had been raised in all states to 21 (Wyoming was the last to do so),[84] and to stricter drunk driving laws, including the institution of sobriety checkpoints.

But drunk driving still kills about 17,000 Americans a year.[85] That is equivalent to a large commercial jet liner crashing each and every week and killing everyone on board. A 1994 survey of 3,000 teenagers listed in *Who's Who Among American High School Students* found that even among these elite youngsters one in five admitted to having driven while drunk—a threefold increase from a comparable group surveyed in 1971.[86] Obviously, the designated driver campaign and other efforts have saved many lives but have not been sufficient to eliminate the public health menace of drivers who are under the influence of alcohol.

Drunk driving, of course, is not the only cause of traffic fatalities—though about 40 percent of all fatal crashes are alcohol related.[87] Other measures have been taken in the past few decades to improve highway safety, including promotion of seat belts; reduction of speed limits (which had been raised back to 65 miles per hour in all but five states by September 1995);[88] mandated child car seats; and redesign of automobiles, including airbags, better side-impact protection, and improved steering assemblies. Highways themselves have been redesigned for safety, with guardrails, median barriers, wider lanes and shoulders, skid-resistant pavements, and improved railway-highway crossings. The result has been a dramatic improvement in driving safety in the United States, from 5.5 deaths per 100 million miles traveled in 1966 to just 1.7 deaths per 100 million miles traveled in 1994.[89] It is estimated that more than 68,000 deaths were avoided between 1982 and 1994 by air bags, child restraints, and increased use of seat belts.[90]

One of the most valuable lessons of the designated driver campaign is that it demonstrated a new way to bring a significant public health message to national awareness. For the first time in history, dialogue in Hollywood entertainment was used as a deliberate strategy of health promotion in an initiative that mobilized large sectors of the entertainment industry.

Winsten has tried to use the same approach to change behavior that accounts for the rise in violence in the United States. In a campaign similar to the designated driver campaign, he began an effort in 1993 to introduce a new idea into television and movie scripts and, ultimately, into ordinary street behavior. When a fight is brewing, a character will raise hands in a timeout gesture and say, "Squash it!"

The phrase "squash it" is an outgrowth of focus groups that Winsten assembled of inner-city teenagers. He found that it was already a street expression that some youngsters understood; his job was to make it cool to use the expression and to follow the behavior that it implied—to walk away from a fight. Winsten used his old Hollywood contacts to insert the expression into TV shows typically watched by minority kids, who are the most frequent victims and offenders in cases of youthful violence. Still, as Winsten explains, there is an enormous difference between preventing deaths from drunk driving and reducing the use of guns:

> With drunk driving prevention, we were dealing with unintentional injuries. Nobody ever set out to cause a drunk-driving crash, to hurt someone. Now we're dealing with intentional injuries with powerful emotions behind them. People are setting out intentionally to do severe bodily harm to another human being.[91]

Many observers are optimistic that the "Squash It!" campaign will ultimately make a dent in the rising tide of violence. As columnist Ellen Goodman wrote in *The Boston Globe*, it can at the very least provide "a wedge, a way to intervene in violence without being overwhelmed by all the problems of urban America. One way to begin turning it around."[92]

Another dramatic demonstration of the link between behavior and disease has come in the form of AIDS. Although AIDS is an infectious disease caused by a virus, its transmission depends on behavior. The AIDS virus, HIV, is not like a cold virus that passes through the air or a yellow fever virus transmitted by a mosquito bite. People are exposed to HIV by engaging in activities that put them into contact with it—most prominently, by sharing needles used for intravenous drugs, by having unprotected sex with people who share needles, or by having unprotected sex with people who have already been exposed to HIV through other sexual partners. The particular pattern of the AIDS epidemic has led public health officials to estimate that if sexual practices

©1994 President & Fellows, Harvard College   SIG PURWIN

and drug-taking behavior were changed—not eliminated, but just changed to make it less likely that HIV could be transmitted—the AIDS epidemic would be markedly diminished.

AIDS: When Risky Behavior Leads to Fatal Disease

When the AIDS epidemic was first recognized in 1982, it seemed to be confined to gay men, especially gay men in large cities who had many sexual partners. Efforts to change behavior were targeted specifically at this group—with rather impressive results. In 1982, for every 100 gay men in San Francisco, 18 were newly infected with HIV; just three years later the rate of new infections had dropped to less than one in 100. This reduction was a triumph of public health education—at least education aimed at a specific, literate, motivated population—and a vivid demonstration of its ability to change intimate behavior. A wide range of efforts, from xeroxed pamphlets passed out on street corners to a massive mailing from the U.S. surgeon general's office to every household in the United States, had focused on getting the word out to all Americans that the only way to stem the spread of AIDS—short of sexual abstinence or mutual monogamy—was by using condoms and by refraining from sharing needles.

161

But the good news turned sour by the second decade of the epidemic, when many gay men, especially those too young to remember the horrors of the early 1980s, started engaging in unprotected anal intercourse again. As one journalist put it, these men said they were "numb with loss, fatalistic about their own survival, unwilling to face a measure of sexual deprivation and eager for the attention showered on the sick and the dying."[93] As a result, the infection rate for gay men started creeping back up; by late 1993, for every 100 gay men under age 25 in the San Francisco area, there were four new infections. For gay men in the thick of an epidemic that just will not go away, the prospect of a lifetime of circumscribed sex can seem too much to bear, especially when half the men in the city are infected with HIV anyway and many of their friends and lovers have already died. Lack of information is not the problem; lack of motivation is. One man who worked in an AIDS clinic stopped using condoms and became infected, even though his work gave him plenty of opportunity to see the disease's effects and its modes of transmission. "So, I'll live to be 55 or 60," said the man, who was 46. "That's not what I had in mind, but it's better than another 30 years worrying about this and living without sexual intimacy."[94]

This same sense of fatalism has affected many heterosexuals as well, especially those living in the inner-city communities now plagued by AIDS. AIDS is the fastest-growing new disease in history. Again, this is the case not because of lack of information but because of lack of motivation or lack of resources to do anything to reduce the risk. Many of the people who constitute the fastest-growing risk group for AIDS in the 1990s—young people between the ages of 13 and 21—have been barraged with AIDS information since grammar school. They know the importance of safe sex and clean needles; the problem is that they don't especially care. "It is increasingly evident that if you are trying to get people to change behavior, giving out information is not enough," says Jonathan Mann of the Harvard School of Public Health, who was the founding director of the Global Programme on AIDS of the World Health Organization (WHO).

> Someone said to me five or six years ago that if you wanted African-Americans in the inner city to use condoms to prevent AIDS, you have to start by helping to assure them of a future. . . . If you don't see a future for yourself, why would you trade off an immediate pleasure to prevent HIV infection, which, if you are infected, may take 10 years to express itself as AIDS? Such a decision requires a long-term perspective.[95]

The major risk factor for AIDS in the United States is "survival sex," according to Karen Hein, former director of the Adolescent AIDS Program at the Montefiore Medical Center in the Bronx. By survival sex she means trading sex for food, shelter, or drugs. Lectures about abstinence, monogamy, and condom use do not hold much credibility among these young people, who use sex as a form of currency.

In other cases, AIDS is spreading because of simple arrogance. "Teenagers have the highest rate of STDs [sexually transmitted diseases] of any age group," says Hein. "It's true for gonorrhea, syphilis, and chlamydia. Why is it that people don't see that it will soon be true for HIV?" According to Hein, the rate of HIV-positive adolescents rose by 77 percent in just two years (from 1991 to 1993)—and half of the transmission was attributable to heterosexual intercourse.[96]

Globally, the AIDS pandemic has also become more and more a disease of the young. "The new face of AIDS worldwide," says Hein, "is the face of teenaged girls."[97] Between its first description in 1981 and the last count in 1994, AIDS had spread to 119 million individuals around the world. Since the beginning of the pandemic 30 million people have been HIV-infected.[98] According to Michael Merson, Mann's successor at WHO, one-half of all HIV infections worldwide have occurred in youth between the ages of 15 and 24.

Public health experts at the Harvard AIDS Institute and elsewhere are now working on the second wave of the pandemic, developing ways to control the outbreak in social and economic contexts quite different from those of the United States. In many AIDS-plagued countries, for example, women's low status, lack of education, and economic dependence on men are so engrained that AIDS is easily spread despite educational programs. When women and girls are dragged into prostitution, and when husbands sleep with many women and then refuse to wear condoms when having sex with their wives, women in many countries have very little power to counter these dangerous practices. The result is that in some places, such as Thailand, where teenage prostitutes are common, those with the highest rate of AIDS are females between the ages of 15 and 19.[99]

The rapid heterosexual transmission of AIDS in Thailand and some parts of Africa may be related as much to the biological characteristics of the AIDS virus as to the social conditions of those countries. Investigators at the Harvard AIDS Institute, led by the institute's chairman, Max Essex, have grouped HIV-1 into at least six different subtypes and have found that the strain most prevalent in Asia and Africa, known as subtype E, grows more efficiently than other subtypes in the female genital tract. As we saw in Chapter 2, Subtype E—which is quite rare in the United States and Western Europe—is the most likely subtype to be spread by heterosexual sex.[100] Unraveling the story of how AIDS spreads differently in different countries demonstrates the importance of collaborations among behavioral, epidemiological, and biological researchers—collaborations that characterize the interdisciplinary approach that is part of any good public health strategy of investigation.

When someone engages in unprotected sex, he or she is using an informal internal calculator to decide whether the risk of getting AIDS or some other sexually transmitted disease outweighs the benefits—money or drugs in some cases and love, acceptance, status, or pleasure in others. Anyone who engages in any behavior thought to be "risky"—whether it is obviously risky like speeding or smoking or risky only in the long term, like eating a hot fudge sundae—is making this internal calculation of risks and benefits. A new goal of public health professionals is to formalize the calculation, which is the work of people in the emerging discipline known as risk assessment. Their mission is to describe, and if possible to rank, the risks people encounter in a way that helps them make meaningful decisions about how to live their lives.

The Science of Risk Assessment

When Congress was considering the 1990 amendments to the Clean Air Act, a group of environmentalists pushed for an aggressive new clause. They wanted a requirement that the added lifetime cancer risk from the emissions of industrial plants, as measured by a hypothetical "maximally exposed individual," be less than one in a million. In other words, each plant would have to prove that it caused no more than one added cancer case for every million exposed individuals in order to continue to operate. Sen. Daniel Patrick Moynihan (D-NY) was uncomfortable with this wording. What, he wondered, did "one in a million" really mean?

Moynihan called in John Graham of the Harvard School of Public Health to give a briefing to members of Congress and their staff about the magnitude of such a number. As Graham recalls:

> Even extremely bright, well-educated people—a lot of these staff people were Harvard and Yale law graduates who had worked on environmental issues for a long time—were perfectly comfortable about writing this number into the law without a real sense of either how the number is calculated or how big this risk is compared to other risks that they encounter in their daily life.[101]

During his briefing, Graham put "one in a million" in context. Using current mortality rates, he said, a baby born in 1990 had a lifetime risk that was higher than one in a million—more on the order of four in a million—of ultimately being killed by being struck on the ground by a crashing airplane.

> No one has seriously suggested that we ought to build all our schools underground, or hold all our business meetings and conferences underground, in order to provide an extra margin of protection. Nor had anyone seriously suggested that we should ban all airplanes that don't meet a one in a million standard.[102]

After Graham's testimony, Congress chose to reconsider the one-in-a-million provision and ultimately allowed more flexibility in regulations. This experience led Graham to the conclusion that opinion leaders need a way to differentiate among "the big risks, the modest risks,

John Graham

and the phantom risks." Thus was born the Center for Risk Analysis at the Harvard School of Public Health, which Graham now directs. Among the questions the center investigates are: Which risks will the public sit up and takes notice of and why? The hazards that push our collective "hot buttons," in Graham's analysis, are

> mysterious, unfamiliar, hard to explain, ones that we perceive are imposed upon us without our knowledge or consent—and that affect, frankly, upper-middle-class white people. If things hit those buttons, then the media gets very interested in them, politicians get very interested, and you can get an enormous amount of attention. Other risks may be factors of 10 and 100 times bigger, but if they don't hit those hot buttons, apathy may ensue. For the major risks, it's often difficult to get the public concerned about them.[103]

In the past decade, public hysteria rose and fell over what experts generally considered to be relatively minimal health risks: Alar in apples, silicone breast implants, salmonella in eggs, electromagnetic radiation, cellular telephones, toxic waste dumps, radiation from nuclear power plants, and asbestos in school buildings. But despite the frenzy over these externally imposed risks, the biggest threat to the public's health remains the things that very few people get worked up about: individual choices. Personal decisions about smoking, eating, driving, and exercising are probably the most significant determinants of health. Smoking alone, as we have seen, is thought to account for one-half of all premature deaths in the United States. Yet when people are asked what they are most worried about in terms of threats to their own and their communities' health, they never talk about individual behavior. The sense that something is being done to them without their consent disturbs people far more than a bigger risk that they are undertaking of their own volition. "The amount of control felt by a person greatly changes the perception of risk," says William Foege of the Carter Center in Atlanta.

> People will accept high risks with cigarette smoking, fast driving, drug use, and other similar activities if they have control of placing themselves at risk. But they will reject even small or nonexistent risks, such as with food additives, fluoridation, or radiation, if they feel they have no control over the exposure.[104]

Graham points to the significant mismatch between the public's perception of the most important threats to public health and the epidemiological evidence:

> If you ask people "What is the most important cause of serious health problems among people? Is it things in the environment, is it people's personal habits, or is it a combination?", the responses you get are consistently that one-third of the people say it is things in the environment, one-third say it is people's personal habits, and one-third say it is a combination. Now, when you get a survey result of one-third, one-third, and one-third, you have a hint that people don't have a clue.[105]

Part of the problem is one of definition. "Risk for the experts means how many people will die," says Peter Sandman of Rutgers University, "but risk for the public means that plus a great deal more. Is it fair or unfair? Is it voluntary or coerced? Is it familiar or high-tech and exotic?"[106]

Sandman was involved in a dramatic demonstration of this incongruity in 1986, when many scientists were blindsided by the public's outrage over a risk they thought was relatively minor. As we saw in Chapter 3, many New Jersey residents had not bothered to have their houses tested for radon, which was considered a major threat by public health officials. Several months later the same residents did finally get worked up about radiation, though it was a source of radiation that state health officials had not even informed them of, since they thought the health risks were so small. To get rid of waste from an old luminescent paint factory in the southern part of the state, which contained small amounts of radioactivity, officials planned to mix the contaminated soil with some regular dirt and to dump it in an abandoned quarry in an uninhabited rural area. But they misread the public's mood. After failing to incite people to action over radon in their homes, state officials were stunned to find that their proposal to dump radioactive dirt met aggressive resistance, which began with a town meeting of 3,000, grew to a rally of 10,000, and ended with a demonstration at the governor's mansion and threats of civil disobedience.

The officials' surprise came in part from their failure to recognize one of the cardinal rules of risk assessment: there has to be a bad guy. "Geological radon had no villain. It's God's radon," says Sandman. "The landfill radon [had] a readily identified villain, which the community felt was unfairly imposing a risk without even telling it, much less asking its permission."[107] Until public health experts learn the significance of public perception in recognizing and controlling health risks, the two camps will continue to talk—and work—at cross purposes.

Public perception is also central to the issue we now turn to: efforts to reanalyze and restructure the health care delivery system. In the next chapter we will look at how the public, in the form of the political process, helps define and shape that system and how health care policy and management have evolved into a crucial component of the mission of public health.

Notes

1. Fredrick J. Stare, *Adventures in Nutrition.* Hanover, Mass.: Christopher Publishing House, 1991, p. 57. See L. H. Kushi, F. J. Stare et al., "Diet and mortality from coronary heart disease: The Ireland-Boston diet-heart study," *The New England Journal of Medicine,* 1985, vol. 312, pp. 811–818.

2. J. M. McGinnis and W. H. Foege, "Actual causes of death in the United States," *Journal of the American Medical Association,* 1993, vol. 270, p. 2207; also, Spencer Rich, "What coroner reports are really telling us," *The Washington Post,* June 15, 1995, p. A19.

3. Edward B. Diethrich and Carol Cohan, *Women and Heart Disease.* New York: Times Books, 1992, p. 9.

4. Walter Willett, personal interview, Feb. 1996.

5. Franz H. Masserli, "This day 50 years ago," *The New England Journal of Medicine,* April 13, 1995, vol. 332, p. 1038.

6. Ibid., quoting from Paul Dudley White, *Heart Disease,* 2nd ed. (New York: Macmillan, 1937, p. 326).

7. Roger B. Spaulding, "Notes on public relations program, Harvard School of Public Health," memo dated Jan. 31, 1951. From the papers of James S. Stevens, Rare Books Collection, Countway Medical Library, Harvard University. Spaulding mentions that Hooton was also noted for "measuring the skull of 'Gargantua,' the French wrestler 'The Angel,' and colleagues on the Harvard faculty."

8. Jean Alonzo Curran, *Founders of the Harvard School of Public Health: With Biographical Notes, 1909–1946.* New York: Josiah Macy, Jr., Foundation, 1970, p. 129.

9. While the homogeneity of the study population made analysis easier, since most potentially confounding factors could be held constant, it did limit the external validity of the study. That is, the findings of the Framingham Heart Study cannot easily be applied to populations that do not look like the original Framingham cohort.

10. Diethrich and Cohan, *Women and Heart Disease,* p. 10.

11. Tom Dawber, personal interview, May 1995.

12. Ibid.

13. "The 'Granddaddy' of heart-disease studies," *Prevention,* July 1986, vol. 38, p. 48.

14. Dawber, personal interview.

15. D. J. Hunter, J. S. Morris, and M. J. Stampfer, et al., "A prospective study of selenium status and breast cancer risk," *Journal of the American Medical Association,* 1990, vol. 264, pp. 1128–1131.

16. Graham A. Colditz, Walter C. Willett, and David J. Hunter, et al., "Family history,

age, and risk of breast cancer: prospective data from the Nurses' Health Study," *Journal of the American Medical Association*, 1993, vol. 270, pp. 338–343.

17. Eliot Marshall, "Search for a killer: focus shifts from fat to hormones," *Science*, 1993, vol. 259, p. 620.

18. "Food and feminine hearts," *Prevention*, April 1992, vol. 44, pp. 9–10.

19. Meir J. Stampfer, Graham A. Colditz, and Walter C. Willett, et al., "A prospective study of moderate alcohol consumption and the risk of coronary disease and stroke in women," *The New England Journal of Medicine*, 1988, vol. 319, pp. 267–273.

20. Meir J. Stampfer, Graham A. Colditz, and Walter C. Willett, et al., "Postmenopausal estrogen therapy and cardiovascular disease: ten-year follow-up from the Nurses' Health Study," *The New England Journal of Medicine*, 1991, vol. 325, pp. 756–762.

21. JoAnn E. Manson, Walter C. Willett, and Meir J. Stampfer, et al., "Body weight and mortality among women," *The New England Journal of Medicine*, 1995, vol. 333, pp. 677–685. An additional important ongoing study of men and the impact of exercise and body weight on health is the Harvard Alumni Health Study, initiated by Ralph S. Paffenbarger, Jr., principal investigator and professor emeritus at Stanford University and an adjunct professor in the Department of Epidemiology at the Harvard School of Public Health. This is a study of 35,000 Harvard graduates, classes of 1916–1950, that began in 1960. Translating their activity levels into a measure called kilocalories, Paffenbarger showed, among other observations, that men who burned fewer than 2,000 calories a week had a higher risk of heart attack. A recent study by I-Min Lee published in *Journal of the American Medical Association* (1993, vol. 270, pp. 2823–2828) showed that the mortality rate was lowest for lean men, defined as weighing on average 20 percent below the U.S. average.

22. Walter C. Willett, Meir J. Stampfer, and JoAnn E. Manson, et al., "Trans-fatty acid intake in relation to risk of coronary heart disease among women," *Lancet*, 1993, vol. 341, pp. 581–585.

23. Diethrich and Cohan, *Women and Heart Disease*, p. 56.

24. Melva Weber, "Smoke and the female heart," *Vogue*, vol. 178, April 1988, p. 384.

25. Sidney M. Wolfe, *Women's Health Alert*. Reading, Mass.: Addison-Wesley, 1991, p. 133.

26. Francine Goodstein, Charles H. Hennekens, and Graham A. Colditz, et al., "A prospective study of permanent hair dye use and hematopoietic cancer," *Journal of the National Cancer Institute*, Oct. 5, 1994, vol. 86, pp. 1466–1470.

27. The Steering Committee of the Physicians' Health Study Research Group, "Final report from the aspirin component of the ongoing Physicians' Health Study," *The New England Journal of Medicine*, 1989, vol. 321, pp. 129–135.

28. JoAnn E. Manson, Meir J. Stampfer, and Graham A. Colditz, et al., "A prospective study of aspirin use and primary prevention of cardiovascular disease in women," *Journal of the American Medical Association*, 1991, vol. 266, pp. 521–527.

29. Walter C. Willett, JoAnn E. Manson, and Meir J. Stampfer, et al., "Weight, weight change, and coronary heart disease in women: risks within the 'normal' weight range," *Journal of the American Medical Association*, 1995, vol. 273, pp. 461–465.

30. Jane E. Brody, "New clues in balancing the risks of hormones after menopause," *The New York Times*, June 15, 1995, p. A20.

31. Diethrich and Cohan, *Women and Heart Disease*, p. 206.

32. Carl E. Bartecchi, Thomas D. MacKenzie, and Robert W. Schrier, "The global tobacco epidemic," *Scientific American*, May 1995, p. 44.

33. Dawber, personal interview.

34. In much the same way, cigarettes are still subsidized by the government for distribution to military personnel who may buy them at post exchanges on military bases.

35. Brian MacMahon, personal interview, May 1994.

36. Fitzhugh Mullan, *Plagues and Politics: The Story of the United States Public Health Service*. New York: Basic Books, 1989, p. 149.

37. Ibid.

38. Ibid., p. 150.

39. Bartecchi, MacKenzie, and Schrier, "The global tobacco epidemic," p. 47.

40. Ibid.

41. Bartecchi et al. found that during the televised coverage of the 1989 Marlboro Grand Prix, for example, the name "Marlboro" was mentioned or shown on the screen nearly 6,000 times, for a total of 46 minutes. "For 18 of those minutes, the Marlboro name was clear and in focus, which represents an estimated $1 million of commercial air time" (p. 47).

42. Trichopoulos and MacMahon's study is described by Kevin Sottak, "Epidemiology," *Harvard Public Health Review*, Spring 1993, pp. 29–35. For the complete study, see D. Trichopoulos, A. Kalandidi, L. Sparos, and B. MacMahon, "Lung cancer and passive smoking," *International Journal of Cancer*, 1981, vol. 27, pp. 1–4.

43. Diethrich and Cohan, *Women and Heart Disease*, p. 56.

44. Ibid.

45. Ibid., p. 48.

46. William Foege, personal interview, Nov. 1993.

47. J. M. McGinnis and W. H. Foege, "Actual causes of death in the United States," *Journal of the American Medical Association*, 1993, vol. 270, p. 2208.

48. ABC "World News Tonight," Oct. 17, 1991, and Feb. 22, 1994.

49. Diethrich and Cohan, *Women and Heart Disease*, p. 50.

50. Bartecchi, MacKenzie, and Schrier, "The global tobacco epidemic," p. 46.

51. Kenneth E. Warner, "Health and economic implications of a tobacco-free society," *Journal of the American Medical Association*, 1987, vol. 258, p. 2080.

52. Questions from the Nurses' Health Study come from the 1990 questionnaire, provided by Walter Willett.

53. Peter Wehrwein, "Food fight: Walter Willett enters the fray," *Harvard Public Health Review*, Fall 1994, p. 36.

54. Eliot Marshall, "Search for a killer," *Science*, 1993, vol. 259, p. 620.

55. Diethrich and Cohan, *Women and Heart Disease*, p. 215.

56. Diethrich and Cohan, p. 241.

57. Greg Gutfeld, "Food and feminine hearts," *Prevention*, April 1992, vol. 44, pp. 9–10.

58. Peter Wehrwein, "Food fight," p. 38.

59. Katherine Griffin, "What's best to spread on your bread," *Health*, Jan./Feb. 1995, p. 30.

60. Walter C. Willett and Alberto Ascherio, "*Trans* fatty acids: are the effects only marginal?" *American Journal of Public Health*, 1994, vol. 84, pp. 722–724.

61. William H. McNeill, *Plagues and Peoples*. Garden City, N.Y.: Anchor Press/Doubleday, 1976, p. 268.

62. Ibid.

63. Fredrick J. Stare, *Adventures in Nutrition*. Hanover, Mass.: Christopher Publishing House, 1991, p. 50.

64. Ibid., p. 51.

65. Fredrick Stare, personal interview, May 1994.

66. Ibid.

67. John Cheever, *The Stories of John Cheever*. New York: Ballantine Books, 1980.

68. John C. Ayres, "An old problem—a new approach," *Harvard Public Health Alumni Bulletin*, May 1952, vol. 9, p. 7.

69. Frederick S. Stinson et al., "Alcohol-related deaths 1993," *Alcohol, Health and Research World*, 1993, vol. 17, pp. 251–260.

70. Rosemary L. McGinn, "For me, alcohol is a disease," *The New York Times*, Dec. 3, 1987, p. A35.

71. Henry Wechsler, Andrea Davenport, and George Dowdall, et al., "Health and

behavioral consequences of binge drinking in college: a national survey of students at 140 campuses," *Journal of the American Medical Association,* 1994, vol. 272, pp. 1672–1677.

72. American Medical Association, news release, Sept. 27, 1994.

73. Walter C. Willett, Meir J. Stampfer, and Graham A. Colditz, et al., "Moderate alcohol consumption and the risk of breast cancer," *The New England Journal of Medicine,* 1987, vol. 316, pp. 1174–1180.

74. Arthur Schatzkin, Robert N. Hoover, and Christine L. Carter, et al., "Alcohol consumption and breast cancer in the epidemiologic follow-up study of the First National Health and Nutrition Examination Survey," *The New England Journal of Medicine,* 1987, vol. 316, pp. 1169–1173.

75. Charles S. Fuchs, Meir J. Stampfer, and Graham A. Colditz, et al., "Alcohol consumption and mortality among women," *The New England Journal of Medicine,* 1995, vol. 332, p. 1249.

76. Jay Winsten, personal interview, May 1994.

77. Ibid.

78. J. D. Podolsky, "Drunk driving goes prime time," *Harvard Public Health Review,* Fall 1989, pp. 23–27.

79. Winsten, personal interview.

80. Ibid.

81. Ibid.

82. Virginia Poole, "Designated driver gets presidential boost," *The Vineyard Gazette,* June 29, 1993, p. 1.

83. National Highway Traffic Safety Administration (NHTSA), *Traffic Safety Facts 1994.* U.S. Department of Transportation, Washington, D.C., Aug. 1995, p. 32.

84. National Highway Traffic Safety Administration (NHTSA), *Effective Dates of State Laws Establishing a Minimum Drinking Age of 21.* U.S. Department of Transportation, Traffic Safety Programs, Washington, D.C., Feb. 2, 1995.

85. NHTSA, *Traffic Safety Facts 1994,* p. 32.

86. Reuters, "More students, in a study, see unsafe schools," *The New York Times,* June 15, 1995, p. A26.

87. NHTSA, *Traffic Safety Facts 1994,* p. 32.

88. National Highway Traffic Safety Administration (NHTSA), *Status of State Speed Limits (65 mph).* U.S. Department of Transportation, Traffic Safety Programs, Washington, D.C., June 21, 1995.

89. NHTSA, *Traffic Safety Facts 1994,* p. 15.

90. National Highway Traffic Safety Administration (NHTSA), *Traffic Safety Facts 1994: Occupant Protection.* U.S. Department of Transportation, Washington, D.C., July 1995, pp. 1 and 4.

91. Ellen Edwards, "Campaign beats up on violence: media experiment urges kids to 'Squash it,' " *The Washington Post,* Dec. 8, 1993, p. C1.

92. Ellen Goodman, "It's not worth it; squash it," *The Boston Sunday Globe,* Dec. 19, 1993.

93. Jane Gross, "Second wave of AIDS feared by officials in San Francisco," *The New York Times,* Dec. 11, 1993, p. 1.

94. Ibid.

95. Ellen Hoffman, "People in peril," *Harvard Public Health Review,* Fall 1994, p. 8.

96. Marsha F. Goldsmith, "'Invisible' epidemic now becoming visible as HIV/AIDS pandemic reaches adolescents," *Journal of the American Medical Association,* 1993, vol. 270, p. 16.

97. Ibid. Karen Hein has since become executive director of the Institute of Medicine, National Academy of Sciences, Washington, D.C.

98. Jonathan M. Mann and Daniel J.M. Tarantola, Eds., *AIDS in the World II: Global Dimensions, Social Roots, and Responses.* New York and Oxford: Oxford University Press, 1996, p. 11.

99. Goldsmith, "'Invisible' epidemic," p. 16.

100. Craig Horowitz, "Has AIDS won?" *New York*, vol. 28, no. 8, Feb. 20, 1995, p. 37.

101. John Graham, personal interview, July 1994.

102. Ibid.

103. Ibid.

104. William H. Foege, "Plagues: perceptions of risk and social responses." In A. Mack, ed., *In Time of Plague: The History and Social Consequences of Lethal Epidemic Disease.* New York: New York University Press, 1991, p. 13.

105. Graham, personal interview.

106. Cristine Russell, "Risk vs. reality: how the public perceives health hazards," *The Washington Post,* Health section, June 14, 1988, pp. 14–18. The article quotes Sandman as saying that risk can be thought of as the result of two equally important factors: hazard and outrage. The public, he says, ignores hazard and focuses primarily on outrage; the experts ignore outrage and focus almost exclusively on hazard.

107. Ibid.

*Providing
Guidance for
Individual
Behavior*

6

Rethinking the
Health Care System

The American health care system spends more per capita than almost any other on earth, yet it offers little in the way of improved health outcomes. Over the past generation, government officials and members of Congress have continually tinkered with health care reform, but so far no one solution has emerged as the preferred way to ensure greater benefits to more people at an affordable cost.

In the 1950s and 1960s the most prominent health concerns of lawmakers were to expand the capacity to deliver services, upgrade facilities, and make health insurance widely available. The primary national innovations of the time were Medicare and Medicaid, which allayed concerns about making health care financially accessible to the elderly and the poor. Reformers also concentrated on a perceived shortage of physicians and other health care workers, leading to the expansion of medical schools and the creation of new professions, such as nurse practitioners and physician assistants. Hospital construction began to burgeon after the passage of the Hill-Burton program in 1946.

In the 1970s and 1980s the focus turned increasingly to the promise and limits of medical technology—on coming up with a balance between reining in costs and safeguarding quality of care. This period witnessed several attempts to reduce the expense of hospitalization and high-tech patient care. Also during these decades, there was an effort to implement quality control, practice guidelines, and peer oversight. Wide variations in physicians' practice patterns sometimes produced different health outcomes and always led to slightly varying costs of care. Rather than lobbying for more hospitals or more doctors, reformers advocated the provision of health care services to underserved areas in the countryside and in the nation's inner cities.

By the early 1990s there was a widespread surge of enthusiasm for a major overhaul of the system. That enthusiasm evaporated by 1995 for reasons that are still being hotly debated. Thus, in the main, the post–World War II history of health care reform has been characterized by incrementalism and a lack of any kind of sustained, broad-based political will for major reform.[1]

The underlying reasons can be traced in part to the American health care divide, captured vividly in early 1993 in a public opinion poll conducted by Robert Blendon of the Harvard School of Public Health. Blendon's questions revealed a schism in American thinking about health care. On the one hand, more than half of those polled said the health care system needed a "complete overhaul"

Robert Blendon

and another 29 percent said it needed "major changes." On the other hand, three out of four were satisfied with their health insurance situation and the quality of care they received.[2] In other words, Americans tended to like their individual health care and to dislike their health care system.

Of course, early 1993 was a time when health care reform was uppermost on the American political agenda, and any latent discontent with the system was likely to express itself. At that time, the gap between personal experience with health care and outlook on the system may have been especially stark and wide. But the similar splits that have shown up at other times and in other polls reflect the fact that Americans for years have had a distinct ambivalence about their own health care system.

This split vision has quite naturally had an impact on public health's role in American health care policy and debate. The interest of public health in systematic approaches to health care, from access to financing, dates back to John Snow and the cholera outbreak in London. But given the lack of American receptivity to big systemwide changes, public health's primary contribution has been to show that various aspects of health care—from the cost effectiveness of a screening program to the rate of negligent injuries in the hospital system—could be quantified and evaluated.

Similarly, the development of design and analysis standards for clinical trials has infused medicine with some measure of objectivity. Though Americans seem remote from any consensus on how best to organize and deliver health care, public health has already played an important role in shaping the system, effectively pushing it to behave more rationally by exposing it to objective analysis and research. Biostatistics and epidemiology, along with decision sciences and evaluation research, guide the public health approach to the health care system. And that approach has served as an important counterweight to the ideology, pressure group politics, and vagaries of public opinion that have colored so many reform efforts.

This chapter looks at some of the ways that public health is helping to evaluate and improve our nation's health care system. We will look at health care decision making, especially the bedrock of all good decisions: reliable information. Public health experts decide what is effective by investigations in clinical research, such as the randomized, double-blind, controlled clinical trial. They also use such methodological innovations as meta-analysis, in which findings from many smaller clinical trials are pooled together to get a better sense of the direction in which the best research is pointing.

Next we look at issues of cost, especially the matter of cost effectiveness: the weighing of a particular intervention's cost, in terms of both dollars spent and health risks encountered, against its expected benefit. This technique is used by public health analysts to guide social choices in a way similar to the risk-benefit calculation an individual might make when considering which treatment to take for an illness.

Going beyond cost effectiveness, we will address one of the most basic questions of health care reformers: how much should the doctor be paid? This will take us on a tour of the resource-based relative value scale, a new mechanism for Medicare and Medicaid reimbursement that has set the standard for federal reimbursement policies. Next we explore quality of care measurement and management through such mechanisms as professional standards review organizations, utilization review programs, peer review organizations, medical malpractice legislation, and the study of errors in medical practice. These methods form the heart of the relatively new science of quality measurement, in which public health professionals have taken a leading role. Finally, we consider health care reform, a topic that periodically rises to and falls from the height of the public policy agenda and flounders again and again in part because of something we observed at the beginning of this chapter: the American public's own ambivalence about what kind of health care system would be ideal.

We begin our investigation by looking briefly at the sometimes rocky relationship between health care delivery and the emerging profession of public health. Over the course of the past century, analysis of the American health care system sometimes seemed a central component of public health; at other times, health care delivery was considered the exclusive domain of physicians and hospitals.

Public Health and Health Care Delivery: A Complex Relationship

The early part of the century was clearly a period when public health professionals took an active role in reforming health care delivery. Alice Hamilton and her contemporaries documented the grievous occupational health hazards and diseases brought about by a rapidly industrializing economy. At the same time, workers' compensation laws made worker accidents and health the responsibility of the business community. The field of industrial medicine blossomed, and industrial health clinics were established.[3] In the 1920s public health officials were among the physicians and economists who served on the privately funded Committee on the Costs of Medical Care, and Roger I. Lee from the Harvard School of Public Health coauthored an influential committee report on medical needs.

Yet for a whole set of complex and interrelated reasons, public health's direct and explicit role in health care receded in subsequent decades. Organized medicine wielded great power and protected health care as its turf. The rapid postwar expansion of employer-based health insurance blunted interest in wider societal health issues. The dominant investment in health research also deemphasized community approaches in favor of the biological sciences; and hospitals, medical schools, and university science departments competed alongside public health schools and institutions.

Before World War II, a trickle of federal research dollars had gone to in-the-trenches study of infectious disease epidemics at the old U.S. Public Health Service's Hygienic Laboratory in Washington, D.C. In the postwar period, however, federal spending on biomedical treatment-oriented research exploded. In 1950, for example, the budget for the National Institutes of Health (NIH), the federal government's leading biomedical research agency, was $46.3 million. It grew almost tenfold, to $400 million, in the next decade. According to medical historian and sociologist Paul Starr, it was the success of the 1954 Salk vaccine trials that fueled this huge burst of political and social support for the NIH and biomedical science. "The magic of science and money had worked" for polio, he wrote, and Americans were hoping the magic could work again for leading killers such as cancer, arthritis, and heart disease.[4]

But if the juggernaut of curative medicine seemed to overwhelm public health concerns in the decades following World War II, it did not completely vanquish them. In some cases, longstanding public health issues gained a more medicalized spin. Community medicine, for example, was a movement to address health care at a local level—a public health kind of endeavor. A 1960 graduate of the Harvard School

176

of Public Health who joined the faculty in 1964, H. Jack Geiger had failed to persuade clinical leaders at Harvard to establish a community health center. He finally succeeded under the sponsorship of the Tufts University School of Medicine in opening a new community health center in the Boston neighborhood known as Columbia Point.[5] The center was a forerunner for the 100 or so community health centers established as part of the Johnson administration's War on Poverty programs.[6]

Any ambiguity about public health's role in health care started to disappear in the 1970s. Leaders in public health began to consider health care not just as a system apart but also as a key factor in shaping the public's health. Bringing impressive quantitative and analytical skills and training to bear, researchers began to break the medical system apart, analyzing what worked and what did not, the quality and appropriateness of the care delivered by the system, the various incentives at play, and the costs of procedures and their benefits.

These efforts did not occur in a vacuum. There was a growing audience for objective critical assessments of the medical system as concern over the rising costs of medical care grew. Increasingly, taxpayers were picking up the bill because the Medicare and Medicaid programs had turned the federal government into the nation's largest buyer of medical services. Critical judgments had to come from outside the clinical practice community, noted Blendon: "You are not going to find many peace studies being done at an Institute of War Planning."[7] At the same time, a more expansive view of public health was also necessary to take on the very complex issues of health care—issues often enmeshed in economic, political, financial, legal, and management problems foreign to old-style public health.

As important as they were to policy planners and various health sector interests, the organization and high cost of America's health care system were largely viewed by the public as technical issues until the 1990s. Then politicians, the press, government officials, and the public clamored for answers to questions about the ins and outs of health insurance, the inner workings of the health care system, and the things that might be done to improve both. This burst of activity, attention, and politics had the effect of pushing public health and its analyses and critiques into the limelight.

More recently, with the failure of a sweeping, government-led reform effort by the Clinton administration, the American health care system is going through a more diffuse transformation, guided by businesses' concerns over cost containment rather than social welfare concerns over access to health insurance. The rise of health maintenance organizations (HMOs) and growing market pressures on medicine have been abundantly chronicled, often with dismay.[8]

Naturally, the exact role that public health will play in marketplace medicine is still evolving, with perhaps only one real certainty: in a cost-conscious system there is bound to be demand for the expertise, techniques, and analytical approaches developed in public health over the past several decades. Some commentators have suggested that public health's systematic evaluation of health care and the setting of quality assurance standards may be a crucial counterweight to the commercialization of medicine.[9] And the ways in which the evaluation of health care has become most systematic have involved all the tools of epidemiology and biostatistics, most especially the clinical trial.

Learning What Works: Clinical Trials and Meta-Analysis

At the end of World War II, some physicians began to suspect that an important part of their armamentarium was doing more harm than good. They worried that one of the most common forms of anesthesia in use at the time, a gas called halothane, was actually killing people by destroying their livers. But their concern was based on a few cases and anecdotal information. The only way to know for sure was to study the question systematically.

Frederick Mosteller was asked to help design such a study. As head of the Department of Statistics at Harvard University, Mosteller had spent much of his career applying statistical analysis to a range of questions, from business administration to law. In 1948 he was asked to turn his attention to a whole new field: public health. "Our job as statisticians was to lay out the quantitative side of the evaluation of medical practices," he recalls. "This involved cost-benefit analysis and decision theory, ideas that the discipline had not yet incorporated."[10]

Frederick Mosteller

The National Halothane Study, published in 1969, showed that halothane, rather than being a dangerous anesthetic, had a lower or equal death rate when compared to other forms of anesthesia. Whether the gas was damaging to livers could not be discerned from the study because of problems with the reliability of medical information provided the investigators.[11] The National Halothane Study was the first of its kind: a large multicenter evaluation run by and for clinicians, answering questions in a novel way. The investigation created a new configuration of people in collaboration; for the first time, physicians and statisticians worked together as full partners to tackle a complicated health care delivery problem.[12] This new configuration led to the development of biostatistics, one of the two numerical disciplines (the other being epidemiology) of public health.

Until 1977, when Mosteller moved across the Charles River to the Harvard School of Public Health to chair the biostatistics department (increasing its faculty from six to 25 within a few short years), biostatistics was a poor relation in the field of public health. "When I came here almost 20 years ago, the raison d'être of this department was to serve other departments," says Nan Laird, the current department chair.

What turned biostatistics into a rigorous intellectual discipline in its own right was the introduction of large-scale clinical trials. As Laird remembers it:

> These trials, which were collaborative, would collect data in a number of different centers all over the United States and all over the world, so they would have their data collection, data management, and data analysis focused in one coordinating center. I think the creation of data coordinating centers, which are largely run by biostatisticians, created an opportunity for biostatistics to grow as its own separate discipline.[13]

In essence, a clinical trial asks whether a particular medical or social intervention—drug treatment, surgery, dietary change—is useful in treating or preventing a particular condition. The most widely accepted form of the clinical trial, the one that has come to be known as the gold standard in medical research, is the controlled, randomized, double-blind study.

- *Controlled* means that the outcomes of patients in an "experimental" group who receive a new therapy or other intervention are compared with those patients in a "control" group, patients who are given either the previously accepted form of treatment or a sham treatment, known as a placebo.
- *Randomized* means that subjects in the study are assigned randomly to an experimental or control group. This practice avoids the bias that would result if group selections were intentionally made (e.g., the subjects placed in the treatment group might tend to be sicker).

- *Double-blind* means that individuals on both sides of the study—that is, both the subjects and the investigators—do not know, until the study is complete, the groups into which the subjects have been assigned. In this way, subtle forms of bias resulting from someone finding the outcome he or she expected are eliminated.

Nan Laird

Each individual component in such a trial helps assure that the outcome is as objective as possible. A review of 250 randomized controlled trials concluded that "when the treatment allocation was inadequately concealed from study participants and investigators, when it was incompletely reported, and when there was no double-blinding, [the] authors tended to exaggerate the effects of the treatment."[14]

The first randomized, controlled, double-blind clinical trial, in the early 1950s, confirmed the usefulness of streptomycin, a new antibiotic, in the treatment of tuberculosis.[15] As the techniques were refined and such trials became more accepted as the ultimate proof of a new treatment's value, conducting large clinical trials became a sort of cottage industry. In the 1970s and 1980s government studies supported multihospital investigations into some of the largest treatment questions of the day. Clinical trials were used to evaluate new drugs for cancer, hypertension, and high cholesterol; new surgery for blocked coronary arteries, breast cancer, and prostate enlargement; and new therapies for infectious diseases, prevention of premature birth, and back pain. Many of the questions central to physicians have been explored in clinical trials during the past 40 years.

Many of the multicenter trials have investigated questions that doctors thought had already been amply answered through clinical experience, but the gold standard has always held the final word. In some cases the use of the gold standard has turned out to be lucky for patients. One of the most dramatic examples involved the surgical treatment of duodenal ulcers through an operation known as gastric freezing.

In 1962 Owen Wangensteen, a prominent surgeon at the University of Minnesota, developed a technique for treating ulcers without drugs or surgery that he said was 100 percent effective in relieving pain, vomiting, weight loss, and other symptoms. The procedure was simple:

A patient is . . . asked to swallow an empty balloon which has tubes attached. After the balloon has been positioned in the stomach, it is continually irrigated with coolant for about an hour. . . . At the conclusion of the procedure, the machine is turned off but the balloon is left inside for an additional 10 minutes until its flexibility has been restored by warming. Then the balloon is removed.[16]

Within two years, gastric freezing had become a widely accepted alternative for the treatment of duodenal ulcer, which until that time could only be treated through symptom relief or, in the case of severe recurrent hemorrhaging, through surgery, which had a 5 to 10 percent mortality rate and was often unsuccessful anyway.[17] In 1964 an estimated 1,000 freezing machines were in use, and 10,000 to 15,000 patients had been treated; five years later the number of machines had more than doubled. The enthusiasm for gastric freezing was barely tempered by the few studies that had been done on the procedure, some of which had downright discouraging results. The studies were small, though—they ranged in size from 10 to 185 patients—and the conclusions mixed. Part of the reason for the confusion was that there was no consistency from one study to another: the studies differed in the way subjects were chosen, in what the control groups were given, and in how success was measured, whether as suppression of acid secretion, relief of symptoms, or some other indicator.

In 1969 a report was published on a major multicenter trial that had standardized all these issues—and turned up some convincing results. The study involved 160 patients at five medical centers. Half were treated with gastric freezing and half with something that just looked like gastric freezing—the cold tube in the mouth and into the esophagus, the balloon in the stomach—without any coolant. After two years, the investigators concluded that "at no time was there a significant difference in the two groups at any period of follow-up" in all clinical and laboratory observations.[18] The procedure was soon abandoned.

The gastric freezing experience illustrates the merit of clinical trials.[19] Yet major, long-term, multicenter trials are costly and cumbersome and can take years to produce results. For decades, biostatisticians have searched for ways to produce statistically valid research results that would require less time, money, and effort. Recently, they have mastered methods of pooling data from a group of smaller studies in a way that allows them to draw inferences that are almost as compelling as those that can be drawn from giant multicenter trials. This pooling, and the statistical analysis that makes it possible, is known as meta-analysis.[20]

The roots of the technique go back more than 90 years, but it was not until September 1988 that the term was added to the key words used for literature searches by the National Library of Medicine.[21] Initially, meta-analysis was criticized because it seemed "to violate a cardinal scientific prohibition: against adding apples and oranges."[22] But in the past decade statisticians have refined methods that allow them to measure not the raw data but the "effect size," whether the effect is statistically significant or not.[23] As one writer explains it:

> In a typical individual study, the results are subjected to standard statistical tests of significance, which reject effect sizes near zero unless the sample size is very large. Meta-analysis, by contrast, looks at the distribution of *all* effect sizes, significant or not. If they are randomly clustered around zero, meta-analysis suggests that the treatment has no effect. But if they cluster off to one side, meta-analysis shows that something is going on, even if the results are not statistically significant in themselves. In this way, meta-analysis can amplify experimental phenomena that are too small to [be seen] in single experiments, and it can find a consistent pattern in apparently contradictory results.[24]

Biostatisticians have used meta-analysis to answer questions plagued by contradictory results from an assortment of smaller studies. A properly done meta-analysis can render a verdict. As a result, biostatisticians can potentially have a major influence on the kind of health care afforded millions of people. Here are some recent examples of raging debates and the information provided by meta-analysis:

- Should women over 50 get regular mammograms? Meta-analysis says yes.[25]
- Does patient education reduce the chance of premature labor and delivery in high-risk women? Meta-analysis says no.[26]
- Should people over age 65 get flu vaccines? Meta-analysis says yes.[27]
- Does exposure to electromagnetic fields through power lines increase the risk of cancer? Meta-analysis leaves this controversial question still unanswered.[28]

Partly because they require sophisticated knowledge and expertise, meta-analysis and clinical trial research are now worlds unto themselves, the province of biostatisticians, specialized medical investigators, and government officials (chiefly at the U.S. Food and Drug Administration). But as Mosteller points out, the postwar development of clinical trial research has had the effect of lending credibility to other kinds of health care evaluations as well. And the complexity of the design, management, and interpretation of clinical trials has encouraged large clusters of biostatisticians to focus on a single clinical problem area.

One recent illustration of this phenomenon is the Center for Biostatistics in AIDS Research (CBAR), established in 1995 at the Harvard School of Public Health. (Its predecessor group, the Statistical Data Analysis Center, was established at the school with a grant from NIH in 1989.) With a staff of nearly 60 biostatisticians evaluating 300 clinical trials that involve 30,000 adult patients and 5,000 children, CBAR is the nation's leading data analysis center for assessments of antiviral thera-

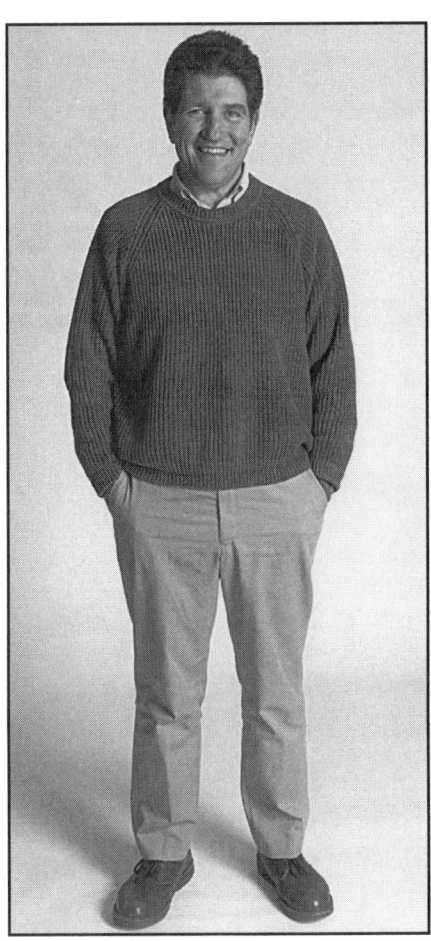

Stephen Lagakos

pies for HIV infection, analyses of treatments for common opportunistic infections, and innovations in the design and conduct of AIDS clinical trials.

Among the influential studies released by the center was one conducted by Stephen Lagakos, CBAR's scientific director, in 1990. It showed that the drug AZT (azidothymidine) was safe and effective for HIV-infected adults who had no symptoms of AIDS but who had low levels of immune system cells known as CD4 cells. More recently, a study of HIV-positive women led by CBAR pediatrics research director Richard Gelber found that giving AZT to pregnant women led to a dramatic reduction in the transmission of HIV from mother to fetus. Gelber's study involved an experimental group of women who received AZT late in pregnancy and during labor and their babies who received AZT for their first six weeks of life. The experimental group had only one-third the rate of HIV transmission as the control group—a one in 12 chance of transmission compared to one in four. When the Harvard researchers analyzed their results midway through the investigation, they called for a halt to the formal research and made AZT available to everyone in the study.[29]

As the tools of epidemiological and biostatistical reasoning have proven to be more and more useful in assessing medicine, health care, and its delivery, the profession of public health has gained entrée into a domain where the authority of the physician had previously been largely unchallenged. As we shall see in the next section, one of these assessment tools—cost-effectiveness analysis—has helped to shape recent developments in the health care delivery system.

The New Science of Cost Effectiveness

During the Great Society of the 1960s, the American philosophy of health care was that more was better.[30] The Medicare and Medicaid programs poured millions of dollars into a health care system that was already prosperous because of abundant government spending on biomedical research and hospital construction. Faith in hospital-based technology was running so high that the Regional Medical Program was created to "encourage the diffusion of new technology to all physicians and hospitals so that all patients might benefit."[31] In the contemporary health care context of hospital mergers, cost consciousness, and suspicions about excessive technology, it is very easy, if facile, to dismiss this spending heyday as errant and excessive. Heather Palmer of the Harvard School of Public Health makes a different argument:

> These policies worked well. There was an enormous growth of health care technology, diffusion of this technology to the public, increased access to technology for the poor. The United States built a health care system that delivered more sophisticated technology to more people than any other in the world.[32]

By the end of the decade, though, the costs of this vast technological marvel started to outrun the results. In 1969, with the Medicare and Medicaid programs mushrooming, President Richard Nixon warned of a "massive crisis" in health care. "Unless action is taken within the next two or three years," he said, "we will have a breakdown in our medical system."[33] At the least, there was an explosion. By 1973 health expenditures had risen as a percentage of the federal budget from 4.4 in 1965 to 11.3.[34]

The business of the 1970s, then, was to find ways to control health care costs while maintaining quality and providing the widest possible access to a technologically advanced system. Public health researchers became an integral part of that business. They established cost effectiveness as one of the new yardsticks for judging an emerging health technology[35] and grappled with ways in which the quality of care might be judged.

A pioneering effort to delve into the cost effectiveness of medical interventions began in 1972, when Howard Hiatt, dean of the Harvard School of Public Health, and emeritus professor Howard Frazier formed the Center for the Analysis of Health Practices. In cooperation with Mosteller and Harvard's Department of Statistics, Hiatt and Frazier assembled more than 100 participants from the Boston area in an interdisciplinary seminar in health and medicine. Seminar participants met for a full evening, first every other week and then weekly, to investigate medical questions by using principles of economics, statistics, law, management, engineering, and biology. Five years later they collected some of their insights into a landmark book, *The Costs, Risks, and Benefits of Surgery*, the first to look at a medical intervention from a cost-effectiveness perspective. As Hiatt wrote in the book's foreword:

> We, in medicine, achieved insights that had previously escaped most of us, not only in analyzing the results of experiments, but in better formulating the questions that needed asking. Further, our colleagues apprised us of previously unappreciated experimental methods and helped

us to a broader understanding of questions asked, and perhaps more often not asked, by our patients. . . . It is not too soon to predict important benefits as well for society and for individual patients.[36]

From these beginnings as a novel transdisciplinary exercise, cost-effectiveness studies have become accepted practice not only in academia and the rarefied world of health research, but in quotidian health care administration. As Milton C. Weinstein of the Harvard School of Public Health wrote recently:

> Cost-effectiveness is a reality, and it is being used. Two countries, Australia and Canada, require cost-effectiveness analyses before authorizing reimbursement of new pharmaceuticals. The larger health maintenance organizations review evidence on cost-effectiveness in designing coverage packages. Companies and universities produce cost-effectiveness analyses for pharmaceutical companies, analyses which are then used to promote products to HMOs, physicians, and government agencies. The U.S. Public Health Service has commissioned a Panel on Cost-Effectiveness in Health and Medicine to develop guidelines for the performance of cost-effectiveness analyses.[37]

At the same time, perhaps because of its intuitive appeal, cost effectiveness analysis has been invoked indiscriminately. Weinstein has editorialized against sloppy semantics, even in medical journals, that labels as cost effectiveness analysis strategies that are really either cost-saving methods or ways to make medicine more effective. Cost effectiveness "is not about costs or cost-cutting," he writes; "rather it is about maximizing the health of populations served within the available resources."[38]

Ironically, one of the targets of this new array of cost-calculating tools has been prevention programs, the bedrock of public health. The case for using sanitation to prevent the mortality and morbidity caused by water-borne infectious diseases, for example, is so obvious to us now that it hardly needs exposition. The notion that a few ounces of prevention is better than pounds of cure has also been used to justify the modern version of this approach: large-scale screening programs of asymptomatic populations.[39] When life expectancy in the United States improved in the 1970s after being stalled for decades, some observers attributed the change to national efforts to screen large numbers of apparently healthy people for heart disease, hypertension, and cancer.

With cost consciousness creeping into the health care system, advocates of aggressive screening programs have broadened their rationale, claiming that screening would ultimately reduce the amount of money spent on health care. But as researchers have taken a closer look at this claim, they have shown that screening programs do not necessarily lower health care expenditures. Indeed, they can often have the opposite effect.

Hypertension is a good example of the hidden cost of screening. What could be cheaper than taking people's blood pressure with a sphygmomanometer? Yet as Louise B. Russell of the Brookings Institution points out:

185

Anyone whose pressure is high on the first test must be retested several times over weeks or months to be sure that the first result was not simply a fluke. If the diagnosis is confirmed, the patient must be evaluated more thoroughly before treatment is prescribed, another expense. Treatment usually involves drugs, which the patient must take regularly for many years to keep his or her blood pressure under control; the cost of drugs averages several hundred dollars a year.[40]

Furthermore, Weinstein's research shows that despite the grave illnesses that hypertension causes (heart attacks, stroke, kidney disease), the costs of prevention are four times higher than the savings in diseases averted for people with moderate or severe hypertension. For people with mild hypertension, the costs are six times higher.[41]

But we would miss the point entirely if we were to conclude that prevention programs, because they cost us something, are a bad use of scarce resources. Clearly, as long as people value life and health, spending some money to preserve them is a worthwhile effort. "Even when prevention does not save money," notes Russell, "it can be a worthwhile investment in better health, and this—not cost saving—is the criterion on which it should be judged."[42]

The more pertinent, economically challenging question, then, is not whether prevention means cost saving; it is which prevention programs provide the most cost effectiveness. In other words, which programs produce the biggest health bang for the buck? To answer this question, public health researchers use as a benchmark the figure of $50,000 per life-year—largely because that is the amount of money that Medicare has deemed worth paying to keep a kidney patient alive for one year on dialysis.[43] The idea, says Thomas H. Lee, an associate professor of medicine at Harvard Medical School, "is that if the government is willing to spend $50,000 to extend the life of a kidney patient for a year, it should be acceptable to spend the same amount to extend the lives of other patients through screenings for other diseases."[44]

It turns out, though, that there is a huge range in the cost effectiveness of screening programs when calculated in terms of saving a single life. With cervical cancer, for example, the net cost of extending life by one year for one woman through a Pap smear every third year for all women ages 20 to 75 would be $13,300. It would jump to $419,000 if the Pap smears were done every other year and to more than $1 million if they were performed annually. Screening for colorectal cancer has turned out to be among the most cost effective: even if all asymptomatic men and women over age 50 were screened through annual tests for hidden blood in the stool, plus periodic sigmoidoscopy, the net cost of extending life by one year for one person would be only $45,000.[45]

The issue of when and how often mammograms should be administered continues to be a subject of debate. To concerns about cost are added those about false positive readings—suspicious findings that lead to biopsies that lead to nothing—and false negatives—results that look normal even though a cancer is brewing. To assess the merit of mammography, society has turned to biostatisticians.

The notion of lead time was developed by George B. Hutchison, a 1960 graduate of the Harvard School of Public Health and a member of its Department of Epidemiology from 1961 until his retirement in 1988. In 1956 and 1957 Hutchison worked in the Division of Research and Statistics at the Hospital Insurance Plan (HIP) of Greater New York, one of the precursors of the HMO movement. At HIP, Hutchison and his co-workers gave mammograms to 30,000 women and compared their health to a comparable group of women who did not receive screening mammograms. The death rate from breast cancer was 40 percent lower for those screened than for those whose disease was detected in some other way. In other words, an earlier diagnosis—or greater lead time—appeared to result in longer survival.

This finding has inspired more routine use of mammograms in screening programs throughout the industrialized world.[46] But there are some unsettling wrinkles. Some researchers following Hutchison have found flaws in the methodology of both his investigation and other studies that have credited mammography with producing longer survival. Moreover, there are general pitfalls difficult for even the most scrupulously designed study to avoid.

One of these is what statisticians call lead-time bias. To explain this concept, let us take the theoretical case of two women with breast cancer. The two women die at the same time. But the first woman, whose cancer was diagnosed earlier through mammography, falsely appears to have a longer survival time than the second woman.

Another potential problem is known as length bias. The most aggressive, fastest-growing tumors are likely to be detected through palpation. Mammography tends to discover slower-growing cancers, which generally have a less lethal course after diagnosis. Thus the survival rates of women whose tumors were detected by mammography will be better than those whose tumors were found by palpation—but only because they had different kinds of tumors in the first place.

These criticisms may call into question the truism that early diagnosis saves lives, especially if preventive screening does not change the course of a disease in the least. But what about other efforts mounted in the name of prevention, for instance, cleaning up the environment to lessen the occurrence of cancer?

In 1994 Tammy Tengs and her colleagues at the Harvard School of Public Health conducted a meta-analysis of the cost effectiveness of various interventions, not only screening programs, undertaken in the name of prevention. Some turned out to be stunningly expensive. The cost of saving one year of life by meeting chloroform emission standards at certain pulp mills, for instance, was $99 billion. For every year of life saved by controlling radiation emissions at nuclear power plants, the cost was $180 million; for a year of life saved by controlling benzene emissions at rubber tire manufacturing plants, $20 billion.

While the control of some environmental and occupational toxins was prohibitively expensive and had negligible likelihood of individual payoff, other preventive techniques were, according to the Harvard researchers, "life-saving bargains": a flammability standard for children's sleepwear; measles, mumps, and rubella immunizations for children; smoking cessation advice for pregnant women; reduced levels of lead in gasoline; and safety standards for concrete construction.[47]

According to Tengs, who is now at Duke University, it would be wrong to conclude "that all societal resources invested in toxin control should be transferred to injury reduction or medicine, or that we should ignore primary prevention entirely and just try to cure disease after it occurs. Rather, efficiency in promoting survival requires that we invest in more cost-effective interventions before proceeding to less cost-effective interventions—in every sector of society."[48]

Inventing a yardstick may seem like a mundane accomplishment. But in health care, as in science, the act of measuring has what is called a Heisenberg effect: it changes the very thing that is being measured. And cost-effectiveness analysis has not been an ivory tower enterprise. As we shall see in the next section, it has been applied to, among other things, a reconsideration of the way in which physicians are reimbursed for their services.

How Much Should the Doctor Be Paid?

By its very nature, cost-effectiveness analysis is grounded in the real world. It asks what a particular service costs, or will cost, and avoids the hypothetical question of what it should cost. In 1988, when Medicare officials asked a group of researchers from the Harvard School of Public Health to devise an alternative reimbursement scheme for doctors, they were asking for something very different from a cost-effectiveness study. Though the researchers were to use many of the same skills, what they were to be doing was analogous to rewriting the rule book rather than simply trying to predict the score.

The original rules had been written in 1965 when, in order to overcome opposition from the American Medical Association, President Lyndon Johnson promised to leave the physicians' fee structure alone in his proposed Medicare legislation. As a result, Medicare paid doctors on a fee-for-service basis and at "customary, prevailing, and reasonable" rates. Such a system turned out to be an invitation to high utilization and high fees for surgical services and other high-tech interventions. The less dazzling side of medicine—general office visits—was left behind. And overall, as we know, the Medicare budget exploded.[49]

The leaders of the Harvard team, William Hsiao and Peter Braun, began with an instrument that was already being used on a pilot basis,

William Hsiao

known as the Resource-Based Relative Value Scale (RBRVS). To put the RBRVS into use nationwide, Hsiao and Braun began by ranking every medical procedure covered by Medicare according to relative value units, or RVUs. The task involved scores of physicans from all over the country who analyzed some 2,000 medical services and arranged them in descending order according to the time and intensity of effort required to perform them. On the RVU scale, for instance, an office visit for a healthy infant received 64 RVUs, and insertion of a pacemaker received 620. Reimbursement was based on a set dollar amount per RVU. In other words, Medicare would pay the surgeon who inserted the pacemaker 10 times more than the pediatrician who conducted the well-baby visit. But while it might seem unfair to pay so much more for an invasive intervention than for cost-effective preventive care, the relative value scale actually reduced the inequities that already existed in the Medicare payment scheme. Under reimbursement standards established for 1986, the Harvard team found that payment for a pacemaker insertion had been 60 times that for a well-baby visit.[50]

For the medical community the RBRVS meant a redistribution of income from Medicare. The typical family practitioner, for instance, would have a 60 to 70 percent rise in income (about $30,000 more per year), whereas the typical thoracic surgeon would have a 40 to 50 percent decline (about $150,000 less per year).

When the RBRVS plan was first announced, the medical community was riveted. In September 1988, *The New England Journal of Medicine* ran a lengthy article by Hsiao, Braun, and their co-workers explaining the system,[51] as well as an editorial by William Roper, director of the Health Care Financing Administration (which had commissioned the study), calling the Hsiao report "pioneering work."[52] A month later the *Journal of the American Medical Association* devoted an entire issue to the RBRVS, including nine articles from the Harvard School of Public Health team and three editorials calling the system an idea whose time had come.[53]

Still, many warnings accompanied the publication of the RBRVS rationale. A chief complaint was that it was based on averages: average time per patient, average total work per condition, average costs per practice. "Have any of us ever met an average physician or an average patient?" asked James Todd, executive director of the American Medical Association, a subcontractor on the Harvard study. Todd, while expressing admiration for the work, withheld an unqualified AMA endorsement.[54] Two members of the Physician Payment Review Commission, which Congress had created in 1986 to advise it on Medicare reimbursement reform, offered similar qualifications. Philip Lee and Paul Ginsburg, writing in the special issue of the *Journal of the American Medical Association*, praised the Hsiao study as "impressive in its magnitude, in its ability to draw on a truly multidisciplinary research team, and in the degree of input by practicing physicians as well as physicians active in health services research." But they cautioned that impressive scholarship did not always translate into good public policy. "Success on these dimensions does not necessarily imply that the results will lead to a fee schedule for Medicare or other payers that is reliable and acceptable."[55]

After the initial flurry of publications and reports, the Harvard team was bombarded with complaints and comments, at the rate of 100 to 200 letters a week. The complaints came mostly from the surgeons and radiologists who would see their incomes go down as a result of the new scale.[56] As the letters kept coming in, more than 20 of the specialty physicians' organizations contracted with outside researchers to try to replicate the Hsiao study. "The story is that we created a new industry of scientists trying to discredit our conclusions," says Hsiao. "I don't believe I know of another piece of research that went through this kind of microscopic examination."[57]

But the initial study withstood the scrutiny as well as the political heat, partly because the RBRVS was appreciated as an attempt to put doctor payment on a rational basis and partly because of an undercurrent of self-interest. As noted previously, a significant number of physicians, particularly in primary care, actually stood to benefit from the new scheme.[58] By 1995 the scale was the reimbursement strategy in place for all of Medicare. In addition, 34 states adopted it for their Medicaid programs, as well as more than 20 Blue Cross/Blue Shield plans and dozens of managed care plans across the country. "I would say the RBRVS probably is the basis, with some modification, in paying for maybe 70 percent of physician services," says Hsiao. "It has been completely accepted by the physician as well as by the payer community."[59]

Public health officials have been called on to evaluate not only cost effectiveness and reimbursement strategies, but also an issue much more difficult to assess: the quality of medical care. Beginning in the 1970s, many of the same factors that led to heightened interest in cost effectiveness also created the new attention given to quality of care: the boom in medical technology, the federal government's emergence as a payer, and the public health researchers' skills at systematic data collection and analysis. In addition, according to Heather Palmer of the Harvard School of Public Health, economists started to argue that American medicine had hit the flat part of the cost-benefit curve: all the extra spending on medical care and application of fancy new technology was not yielding much in the way of improved health for the American people. In the next section, we turn to the emerging question of quantifying quality, which in the past few decades has become a central concern of public health.

Quantifying Quality

There is nothing new about attempts to study and quantify the quality of medical care. Florence Nightingale's studies of the mortality rates of wounded British soldiers during the Crimean War in the late 1850s were, in essence, quality assurance studies. In 1916 E. A. Codman published a study describing the condition of patients one year after surgery. Remarkably, it was the quality of his own care that Codman investigated: the subjects of his study were his surgical patients.[60]

But these early assessments viewed quality of care as purely a medical matter, a question best answered by physicians. The tendency was to focus on patients, not populations; on physician contributions, not the effect of the overall health care system; and on changes in health status as an indicator of quality.[61]

In the 1970s the outlook shifted, and quality-of-care issues were joined with those of cost; they have been inseparable ever since. In simplified terms, public health professionals during this time were evaluating how much medical care people were getting. Fee-for-service medicine was widely characterized as creating an incentive for physicians to overtreat their patients and for patients to overutilize the medical care available to them. Wide geographic variations in the numbers of certain kinds of expensive operations performed, such as cardiac bypass, seemed to point to inappropriate care.

From 1971 to 1988 the RAND Corporation carried out an innovative look at the link between cost and utilization known as the RAND Health Insurance Experiment. Led by Joseph Newhouse,[62] who later joined the faculty of the Harvard School of Public Health, the experiment analyzed data collected between 1974 and 1982 from 2,750 American families on their use of medical services. It was the first randomized trial to assess the impact of cost sharing—requiring patients to pay a portion of their own medial expenses—on quality of care. Using a sophisticated model, the families were assigned either to one of 14 fee-for-service insurance plans or to an HMO.

After eight years, the investigators concluded that full coverage led to more people using more services without any measurable improvement in the average person's health.[63] The researchers also reported that cost sharing had no adverse health effects, or only minimal ones, for the average individual. The only exception was for the sick poor: those in low-income groups whose care was absolutely free (as opposed to those who had to pay for some portion of their care) had a predicted 10 percent decline in mortality rates.[64]

Perhaps as a result of the RAND experiment, Congress postponed proposals in the 1980s to implement cost sharing in the Medicaid program. At the same time, the experiment provided statistical ammunition for critics of free and comprehensive coverage, and employers introduced more cost sharing into their insurance plans. The experiment also established health outcome measures that have been used in many major offspring studies—the Medical Outcomes Study, for example, and others on the appropriateness of care.[65]

A different response to the growing concern about overutilization was the establishment in 1974 of professional standards review organizations (PSROs). Groups of local doctors were given government grants to review Medicare and Medicaid files to track down care that was unnecessary or did not meet local standards. While PSROs were lambasted by consumer groups as foxes guarding the chicken house,[66] their work ultimately led to utilization review, now an important cost-control strategy used by HMOs.

In the 1980s overutilization remained an important issue, particularly as hospitals started to compete vigorously for patients. Various Reagan administration policies were designed to turn the health care delivery system into a free market enterprise. More recently, the emphasis has slowly started to shift from zealotry about overuse to concern that some people are not getting enough medical care or, to put it more precisely, do not have access to the best available care. This shift is largely the product of more Americans enrolling in cost-conscious HMOs and other health plans where the incentive structure is to avoid using expensive services.

Peer review organizations, which succeeded PSROs, are specifically designed to maintain quality standards in this new cost-containment environment.[67] Some state governments have imposed mandates on insurance companies to keep them from cutting certain services for cost reasons. One high-profile issue has been the amount of time a woman can stay in the hospital after giving birth. When insurance companies moved to limit postpartum stays to 24 hours or even less, lawmakers in several states introduced bills to force the insurers to pay for longer hospital stays.

In addition to focusing on these knotty utilization issues, researchers and policymakers involved in quality-of-care matters have looked at ways to prevent medical mistakes. Clearly, a mistake—such as a missed diagnosis or faulty treatment—can lead to a poorer health outcome, robbing medical intervention of any health justification. But a mistake can also cost money and is therefore anathema in these days of cost-conscious health care. As Palmer writes:

> Outright errors in diagnosis and treatment waste resources and may even harm patients so that their injuries require further expensive

192

treatment. Failure to design and operate safe, timely and efficient systems for implementing patient care can cause expensive errors and wasted resources. Therefore, improving the quality of decisionmaking and implementation for health care is an essential component of any cost control program.[68]

For public health researchers, one of the chief challenges of doing research into quality of care is deciding what should be measured. Is quality measured best by patient satisfaction, by adherence to preset practice guidelines, or by morbidity or mortality rates? Is it more important to tally up rates for hospitals, or are doctor-specific rates more germane? Hard-to-resolve philosophical dilemmas also crop up in quality-of-care measurement. If patient satisfaction is given too much weight, quality of care gets defined by what patients demand or expect, regardless of resource and technological constraints.[69]

Despite some truly daunting political, ethical, and methodological problems, measuring medical malpractice, error, and injury has in many ways been more clear-cut than evaluating the quality of health care. Researchers have been able to assemble reliable figures on malpractice and error rates and in doing so put discussion of reforms and prevention efforts on a factual basis. In early 1970s Don Harper Mills of the University of Southern California School of Medicine conducted a path-breaking study of medical malpractice, combing through 20,000 records from California hospitals. Mills and his colleagues found "potentially compensable events" in one out of every 21 hospitalizations. Negligent events—or, in the parlance used in the study, "potentially litigable events"—were more rare, occurring in only one out of every 125 hospitalizations.

The Mills study received a lot of attention, most of it well deserved. But as an epidemiological study of medical injury in hospitals, it had some serious flaws. The records were provided on a voluntary basis, so whether the sample fairly reflected the hospital population was open to doubt. Moreover, whether hospital records could be relied on as accurate measures of untoward or negligent events loomed as a major question.

The Harvard Malpractice Study, its results first published in 1991, took the Mills study as a model, fixed some of its problems, and became recognized as the most reliable study of medical malpractice conducted to date. It grew out of a joint Harvard Law School/Harvard School of Public Health study group on medical malpractice. Formed in 1984, the study group got mired in the general ignorance about how much malpractice actually occurred. The Harvard researchers first approached Massachusetts officials about conducting a systematic study of medical injuries in the state's hospitals, but they got a cool response. A research project was viewed as possibly jeopardizing passage of pending medical malpractice legislation.

The study group fortunately found a willing collaborator in David Axelrod, the forceful New York state health commissioner who was pushing a variety of policies aimed at improving the quality of care in New York hospitals. As part of a malpractice reform package, Axelrod persuaded the New York legislature to spend $4 million on a study of malpractice. With this mandate and this funding, the Harvard researchers set out to do a study of such scope and high methodological standards that it could withstand all scrutiny.[70]

The investigators, led by Howard Hiatt, began with several pilot studies to routinize the reviewing of hospital records and to establish that they were indeed a reliable source of information about medical injuries in hospitals. Their sample was large—31,429 records from 51 hospitals—and randomly selected. To obtain the cooperation of hospitals, the Harvard researchers had to promise that patients would not be tipped off by interviewers if their injury was, in fact, caused by hospital error or negligence. Patients were told only that the study was about "the economic consequences of hospitalizations." The interviewers themselves were similarly uninformed about the true purpose of the study. And the researchers were careful to pick 1984 as their study year because it would put cases of injury past the statute of limitations for a malpractice claim.[71]

When the study results were made public in 1990, it was front-page news in *The New York Times*.[72] The Harvard study found that 3.7 percent of hospitalized patients suffered an "adverse event" and that about one-quarter were due to negligence. The results were consistent with those from California. The most striking may have been the researchers' finding that 13,400 New Yorkers had died as a result of medical treatment. Extrapolated to the nation as a whole, the death toll would be 150,000—three times the number of motor vehicle accident fatalities and 25 times the number of occupation-related deaths.[73]

The Harvard Malpractice Study helped change the terms of the malpractice debate. Although they were careful about adding some important qualifications, the researchers emphasized that their study, combined with Mills's work in California, proved that "modern medical care is an inherently risky enterprise."[74] At the same time, they found that the tort system for dealing with malpractice was more prone to errors than the medical care system. Only about one of every seven cases of negligence generated a lawsuit. And of the people who filed malpractice claims, the researchers found evidence of negligence in only one of six cases. In other words, the vast majority of negligently injured patients were not suing, and the vast majority of those suing were not negligently injured.

The solution the researchers proposed was one of strict liability, similar to workers' compensation, whereby a hospital or medical organization would be responsible for any injury during treatment, regardless of fault. The notion was that strict liability would both rationalize compensation and provide an incentive to prevent injuries. So far, some state-level experiments with no-fault systems in obstetrics have not been

194

Lucian Leape

successful, but the Harvard researchers have argued that those systems were poorly designed and diluted prevention incentives.

Lucian Leape, one of the Harvard malpractice researchers, has studied medical errors. Leape estimates that 180,000 people a year die as a direct or indirect result of physician-caused injury, "the equivalent of three jumbo-jet crashes every two days."[75] The problem, he believes, is not that doctors and nurses make mistakes—since all human beings are imperfect—but that the health care system makes it impossible for them to make mistakes safely. It is almost as if the system itself has embraced the myth that physicians are infallible. "The basic health care system approach is to rely on individuals not to make errors rather than to assume they will," Leape states. "Systems that rely on error-free performance are doomed to fail."[76]

Leape believes that the technology to make certain procedures safer is available and that it has been used with much success in the high-risk specialty of anesthesiology. Why not depend on the memory of a computer, rather than that of the fallible human brain, to keep track of such mundane but error-fraught procedures as medication schedules, patients' drug allergies, and dosing orders? Why not build more redundancy into the system, so that errors can be caught when the same wrong decision is made a second or third time? Why not do away with working conditions that lead to many of the factors—time pressures, fatigue, fear, stress—that contribute to mistakes in the first place? Although these changes would all be useful, according to Leape:

> The most fundamental change that will be needed if hospitals are to make meaningful progress in error reduction is a cultural one. Physicians and nurses need to accept the notion that error is an inevitable accompaniment of the human condition, even among conscientious professionals with high standards. Errors must be accepted as evidence of system flaws, not character flaws.[77]

Making rational informed medical decisions is at the heart of a relatively new discipline in public health known as decision analysis. The essential idea is to help a decisionmaker make the best choice when faced with uncertainty about the costs, risks, and benefits of different options. Such situations arise as frequently in matters of personal health (should I get the operation or not?) as in social policy choices. Decision analysis separately gauges values (how much do I like it?) and probabilities (how likely is it to occur?). Then the technique takes account of both values and probabilities to help guide the decision. Of course, the approach does not eliminate uncertainty, but it does promote systematic thinking in the face of uncertainty.

Decision analysis emerged in the late 1960s and early 1970s as a field of scholarship in business administration. By 1980, just about every top business school in the nation had incorporated decision analysis into the core curriculum for a master's degree. But the field earlier had an uphill battle for recognition, both in business schools and, later, in schools of public health. Howard Raiffa, for instance, who taught at the Harvard Business School, remembers the "howl" of protest sent up by his colleagues when the first courses in decision science were introduced as part of the core curriculum requirements for a master's in business administration. "What are those guys trying to do?" Raiffa recalls the critics complaining. "Decision making is an art, not a science, and you can't automate the mysterious, creative, indescribable insights of the experienced practitioner."[78] Writing in a foreword to *Clinical Decision Analysis* by Milton Weinstein and Harvey Fineberg, Raiffa emphasizes the value of decision analysis not only in business but also, and perhaps especially, in medicine:

> Medical decisions are important enough and repetitive enough for us to try painstakingly to dissect the decision process. In doing this, there are some obvious pluses: we can systematically incorporate objective statistical data; we can elicit and combine the judgments of several experts; we can adaptively and systematically combine the experiences of others; we can calibrate the experts; we can correct for dysfunctional biases; we can elicit the preferences of patients for different medical outcomes.[79]

One of the most critical factors in making any decision in any sphere, of course, is having the right information. After the data are collected, decision scientists can help clarify the steps that are taken to sift through, evaluate, and analyze those data. But without a body of facts to begin with, no decision can ever be made. That is why public health professionals, with their expertise in epidemiology and biostatistics, have been at the forefront in promoting the new science of decision analysis.

They have been in the forefront, too, of one of the most hotly debated issues of the early 1990s, to which we now turn our attention: health care reform.

Health Care Reform: Polls and the Promised Land

A national plan that would provide universal health insurance, and presumably assure access to the health care system itself, has waxed and waned as a public health ideal for over 50 years. Although the American Medical Association (AMA) waged war throughout the 1940s and 1950s against anything that smacked of socialized medicine, the American Public Health Association in 1944 passed a resolution calling for a "national medical care plan" that would make available to the entire population all essential preventive, diagnostic, and curative services.[80] Congressional legislation that would have created a similar plan was introduced annually from 1939 to 1949 but was invariably thwarted by opponents' refusal to hold hearings.[81] With the Red scare and anticommunist politics at a fever pitch, the AMA's warnings about socialized medicine found an audience. Besides, as we have seen, Americans had reason to be satisfied with their health care in the decades following World War II. Employers covered their employees with little fuss and at little or no expense to employees. And jobs were plentiful.

By the early 1990s, however, employers began to cut back on insurance coverage or stopped providing it altogether, and the middle class became anxious.[82] News stories on the plight of people who had lost or had been denied health insurance abounded. A flurry of reports and news accounts made invidious comparisons between American expenditures on health care and those of other countries. In tandem with spending comparisons were comparisons of some basic health measures, such as life expectancy and infant mortality—measures on which the United States often came out behind.

In 1990 public support for national health care reform reached a 40-year high. Polls showed that the majority of Americans believed that health insurance should be available to everyone.[83] In the area of politics, the wake-up call came in 1991, when underdog Harris Wofford, a Pennsylvania Democrat, won election to the U.S. Senate on a platform of health care reform.[84] The following year, presidential candidate Bill Clinton made health care reform a central plank in his platform.

Many people in public health seized this opportunity to talk about long-festering problems with the nation's health care system. They talked about incentives to overtreat in a system based on fees for service and private third-party payers. They emphasized the dollar value of prevention. "Medicine can cure you, but public health will keep you well," said Fernando Trevino, executive director of the American Public Health Association.

In his book, *Your Money or Your Life,* Marc Roberts of the Harvard School of Public Health translated into the vernacular the arcane explanations for the curtailment of health insurance coverage. He likened the private, fee-for-service insurance system, with its incentives to overtreat, to "restaurant insurance," with the unwitting and not directly accountable patron eagerly taking the waiter's advice to order the dry Taittinger champagne and out-of-season strawberries.[85] Roberts compared the way private health insurance is financed today with "trying to sell fire insurance in a town in which 10 percent of the homes are already burning." Indeed, he said, there is relatively little mystery to selling such insurance. Companies "know who's going to be sick next year—it's the people who are already sick."[86] Furthermore, Roberts

197

Marc Roberts

pointed out, the incentives are built in such a way that the insurers try not to cover clients who have the greatest need for health coverage—the sick.

For a while after Bill Clinton's election in 1992, it looked like fundamental health care reform was possible. He and his wife, Hillary Rodham Clinton, devoted the first two years of his presidency to writing and trying to pass the Health Security Act, which would have established a system of "managed competition" between insurance companies grouped into government-run regional health alliances.

Because public opinion played such a key role in putting health care reform on the national political agenda in the early 1990s, assessment of popular beliefs and moods about health care became a central feature of the Clintons' efforts to get their plan enacted into law. Historians such as Paul Starr had chronicled how special interests, particularly organized medicine, and cultural attitudes toward medicine and science have long shaped the health care system. In the 1990s the polling techniques of Robert Blendon both reflected and reenforced the new power of public opinion in the health care debate. Blendon's polls, and his interpretation of them, helped identify the middle-class anxiety about health insurance that launched the Clinton reform effort.

These polls were also later to identify some of the key reasons why the Clinton plan fell out of favor, both after the fact and during its slow tortuous demise in 1993 and the early part of 1994. The public's distrust of government, lack of any real-world example of a "health alliance," and skepticism about the claim that the plan would not ultimately result in a major tax increase were some of the currents of public opinion that Blendon identified. He also hit on the fact that health care, unlike many other issues, is quickly personalized. As the Clinton plan was being debated in the spring of 1994, Blendon commented that Americans, after favoring any kind of reform plan, were finally beginning to ask how the Clinton plan and its competitors would affect them: "They want the question, 'Will my family ultimately be better off?' answered."[87] When a significant number of Americans came to believe that with the Clinton plan the answer would be *no*, it was doomed.

Does this defeat mean health care reform will never happen? Not necessarily. Noting that the issue keeps coming back, in one form or another, Blendon points out that "the important thing to remember is that these are cyclic issues."[88]

Marc Roberts predicts that health care reform will "almost certainly" rise again to the front of the nation's political agenda. What is less certain, he says, is whether the American public will ever reach a consensus on what health care system is best. "The reason is simple if sobering," says Roberts. "There just is no way to give Americans everything they want from their health care system. What we apparently desire is unlimited access to the world's best care, with no organizational or bureaucratic barriers, and without imposing real costs on either ourselves, our government, or the economy. Such a promised land is not attainable."[89]

Notes

1. Barbara Starfield and Lisa Simpson, "Primary care as part of US health services reform," *Journal of the American Medical Association*, 1993, vol. 269, pp. 3136–3139.

2. "A Survey of American Attitudes Toward Health Care Reform," conducted by the Program on Public Opinion and Health Care at the Harvard School of Public Health and by Marttila & Kiley, Inc. The survey was done for the Robert Wood Johnson Foundation, Princeton, N.J. Two thousand people were interviewed between March 18 and March 25, 1993.

3. Rosemary Stevens, *In Sickness and in Wealth: American Hospitals in the Twentieth Century*. New York: Basic Books, p. 83.

4. Paul Starr, *The Social Transformation of American Medicine*. New York: Basic Books, 1982, p. 347.

5. Geiger, who is now chairman of the Department of Social Medicine at the City University of New York School of Medicine, explains the hesitancy among his medical school colleagues as a product of the times. "The year 1964," he explained in a personal communication written more than 30 years later, "was a time when the separation of public health and clinical medicine was still quite rigid, and maintained as such, in a kind of disciplinary purity, by schools of medicine and schools of public health alike. One of the core innovations of community health centers was the attempt to merge the clinical care of individuals with an approach to whole populations and communities as target and partner in a single institution, effectively uniting what had been medicine and public health. This was something that (at the least) seemed a bit strange to both types of schools at the time"(Jack Geiger, personal communication, Jan. 22, 1995).

6. Starr, *Social Transformation*, p. 371.

7. Robert Blendon, personal interview, Mar. 1996.

8. Jerome P. Kassirer, "Managed care and the mortality of the marketplace," *The New England Journal of Medicine*, 1995, vol. 333, pp. 50–52. Kassirer, who as editor of *NEJM* has

great influence, wrote in the editorial, "Market-driven health care creates conflicts that threaten our professionalism. . . . Physicians will be forced to choose between the best interests of their patients and their own economic survival."

9. Leon Eisenberg, "Medicine—molecular, monetary or more than both?", *Journal of the American Medical Association*, 1995, vol. 274, p. 334.

10. Frederick Mosteller, personal interview, Oct. 1993.

11. John P. Bunker, et al. *The National Halothane Study: A Study of the Possible Association Between Halothane Anesthesia and Postoperative Hepatic Necrosis.* Bethesda, Md.: National Institutes of Health, National Institute of General Medical Sciences, 1969, pp. 407, 417–418.

12. Frederick Mosteller's propensity to collaboration, which actually predated the National Halothane Study, was the source of some gentle ribbing from his colleagues at the Harvard School of Public Health. One of them, Donald Berwick, regaled him with an 11-stanza poem called "The Last Collaborator" on the occasion of Mosteller's seventieth birthday in 1986. The poem begins:

> On a high and secret mountain on a South Pacific isle
> Lived a hermit in a mud house in a most reclusive style.
> He had not clothes nor money, neither dishes nor a bed,
> And he had never even written one short monograph with Fred.

13. Nan Laird, personal interview, Nov. 1993.

14. Drummond Rennie, "Reporting randomized controlled trials: an experiment and a call for responses from readers," *Journal of the American Medical Association*, 1995, vol. 273, p. 1054. The recommendation was to guarantee randomness in the assignment of patients to a treatment or a control group and to guarantee that no one knows who is assigned to which group until the study is complete.

15. The story of these earliest clinical trials is told in Samuel Epstein and Beryl Williams in *Miracles from Microbes: The Road to Streptomycin* (New Brunswick, N.J.: Rutgers University Press, 1946).

16. Lillian Lin Miao, "Gastric freezing: an example of the evaluation of medical therapy by randomized clinical trials," in J. R. Bunker, B. A. Barnes, and F. Mosteller, eds., *Costs, Risks, and Benefits of Surgery.* New York: Oxford University Press, 1977, p. 198.

17. Ibid., p. 199.

18. The centers involved were the University of Chicago School of Medicine, Duke University Medical Center, Louisiana State University School of Medicine, Scott and White Clinic, and Vanderbilt University School of Medicine. Study results were published by J. M. Ruffin, J. E. Grizzle, and N. C. Hightower, et al., in "A cooperative double-blind evaluation of gastric 'freezing' in the treatment of duodenal ulcer," *The New England Journal of Medicine*, 1969, vol. 281, p. 16.

19. The gastric freezing story is resurrected periodically whenever controlled clinical trials come under attack on ethical grounds. Many of these studies withhold treatment from at least some patients, even if standard treatment exists, because the belief is that a new therapy is best compared with a placebo—a totally useless pill—than with an older therapy. But the use of placebos for otherwise treatable conditions is unethical, write Kenneth J. Rothman of the Boston University School of Public Health and Karin B. Michels of the Harvard School of Public Health in "The continuing unethical use of placebo controls" (*The New England Journal of Medicine*, 1994, vol. 331, p. 396). "No scientific principle," they write, "requires the comparison in a trial to involve a placebo instead of, or in addition to, an active treatment." Critics have wondered how doctors can justify giving a placebo to a randomly chosen group of people rather than giving them the new therapy from which they are very likely to benefit. "Although withholding an accepted treatment may occasionally seem innocuous," Rothman and Michels continue (p. 397), "allowing investigators to do so runs counter to the ethical principle that every patient, including those in a control group, should receive either the best available treatment or a new treatment thought to be as good or better. Instead, it concedes to individual investigators and to institutional

review boards the right to determine how much discomfort or temporary disability patients should endure for the purpose of research."

20. One useful introduction to meta-analysis can be found in a report by the National Research Council, *Combining Information: Statistical Issues and Opportunity for Research* (Washington, D.C.: National Academy Press, 1992).

21. Meta-analysis was first used in 1904, when a British statistician, Karl Pearson, collected data from several studies of the military's experience with vaccination against intestinal fever. Pearson concluded that the vaccine was ineffective (Graham Colditz, Elisabeth Burdick, and Frederick Mosteller, "Heterogeneity in meta-analysis of data from epidemiologic studies: a commentary," *American Journal of Epidemiology*, 1995, vol. 142, p. 371).

22. Charles C. Mann, "Can meta-analysis make policy?", *Science*, 1994, vol. 266, p. 960.

23. One of the problems with meta-analysis has been what the experts call heterogeneity—that is, differences in study outcomes that cannot be easily explained. Heterogeneity often results from variations in study design, measurements of exposure to be studied, and factors that vary along with the factors being investigated. Statistical methods exist, though, to account for heterogeneity, according to Graham Colditz, Elisabeth Burdick, and Frederick Mosteller of the Harvard School of Public Health. ("Heterogeneity in meta-analysis of data from epidemiologic studies: a commentary," *American Journal of Epidemiology*, 1995, vol. 142, p. 381). They write: "By developing a clear set of methodological guidelines for reviewing prior research and using statistical procedures to summarize the results, meta-analysis offers the potential to make reviewing the literature more of a science than an art."

24. Mann, "Meta-analysis," pp. 960–961.

25. Karla Kerlikowske, Deborah Grady, Susan M. Rubin, Christian Sandrock, and Virginia L. Ernster, "Efficacy of screening mammography—a meta-analysis," *Journal of the American Medical Association*, 1995, vol. 273, pp. 149–153.

26. W. J. Hueston, M. A. Knox, and G. Eilers, et al., "The effectiveness of preterm-birth prevention educational programs for high-risk women: a meta-analysis," *Obstetrics and Gynecology*, 1995, vol. 86, pp. 705–712.

27. P. A. Gross, A. W. Hermogenes, and H. S. Sacks, et al., "The efficacy of influenza vaccine in elderly persons: a meta-analysis and review of the literature," *Annals of Internal Medicine*, 1995, vol. 123, pp. 518–527.

28. M. A. Miller, J. R. Murphy, and T. I. Miller, et al., "Variation in cancer risk estimates for exposure to powerline frequency electromagnetic fields: a meta-analysis comparing EMF measurement methods," *Risk Analysis*, 1995, vol. 15, pp. 281–287.

29. Peter Wehrwein, "AIDS on trial," *Harvard Public Health Review*, Fall 1995, pp. 29–34.

30. R. Heather Palmer, Avedis Donabedian, and Gail J. Povar, *Striving for Quality in Health Care: An Inquiry into Policy and Practice*. Ann Arbor, Mich.: Health Administration Press, 1991, p. 8.

31. Ibid.

32. Ibid.

33. Starr, *Social Transformation*, p. 381.

34. Ibid., p. 399.

35. Milton C. Weinstein, "Spine update, editorial comment," *Spine*, 1996, vol. 21, no. 5, p. 1.

36. Howard H. Hiatt, "Foreword." In J. P. Bunker, B. A. Barnes, and F. Mosteller, eds., *The Costs, Risks, and Benefits of Surgery*, New York: Oxford University Press, 1977, p. x.

37. Weinstein, "Spine update," p. 1.

38. Ibid.

39. The idea behind mass screening programs is not necessarily to prevent diseases but to detect them early enough so that treatment will be more effective, perhaps even curative.

40. Louise B. Russell, *Is Prevention Better Than Cure?* Washington, D.C.: Brookings Institution, 1986, p. 4.

Rethinking the Health Care System

41. Milton C. Weinstein and William B. Stason, *Hypertension: A Policy Perspective.* Cambridge, Mass.: Harvard University Press, 1976, p. 60.

42. Russell, *Is Prevention Better Than Cure?*, p. 5.

43. The End-Stage Renal Disease Program is the only one that guarantees federal coverage for all costs incurred for a particular chronic condition. Its cost has grown beyond all projections, in part as a direct result of the program's success. When the legislation was passed in 1972, people with end-stage renal disease generally died within six months of diagnosis. Based on that life span, legislators projected that about 90,000 Americans would need kidney dialysis by the year 2000. But once dialysis became widely available, dialysis centers blossomed under the reimbursement policies of the federal government, and people with kidney failure were able to live relatively normal lives on dialysis, some for 20 years or more. In 1993 about 165,000 Americans were on dialysis—nearly twice the number expected to be on dialysis by the turn of the century. By the time the year 2000 arrives, according to the latest estimates, the number is likely to just about double again, to perhaps as many as 300,000 patients on dialysis (John K. Iglehart, "The American health care system: the End-Stage Renal Disease Program," *The New England Journal of Medicine*, 1993, vol. 328, p. 368).

44. Spencer Rich, "A costly ounce of prevention: early medical screening may save lives, but not money," *The Washington Post*, Aug. 29, 1993, p. C3.

45. Ibid., p. 3.

46 "Retirements: George Barkley Hutchinson," *Harvard School of Public Health Alumni Bulletin*, Oct. 1988, pp. 33–34.

47. Tammy O. Tengs, et al., "Five hundred life-saving interventions and their cost-effectiveness," *Risk Analysis*, 1995, vol. 15, pp. 369–390.

48. Tammy O. Tengs, et al. "Five hundred life-saving interventions and their cost-effectiveness," manuscript submitted for publication, July 25, 1994, pp. 10–11.

49. Almost as soon as President Johnson handed the first Medicare cards to Harry and Bess Truman (Howard Fineman, "Mediscare," *Newsweek*, Sept. 18, 1995, p. 38), costs started escalating, and Congress set about to control runaway expenditures. As E. Richard Brown of the University of California at Los Angeles School of Public Health put it: "Cost containment competed with and then replaced the original legislative goal of equity as the dominant concern of state legislatures as well as of the Congress" (E. Richard Brown, "Medicare and Medicaid: band-aids for the old and poor," in V. W. Sidel and R. Sidel, eds., *Reforming Medicine: Lessons of the Last Quarter Century*, New York: Bantam Books, 1984, p. 51).

50. "Study would redistribute physicians' income," *Health Sciences Report: Harvard School of Public Health*, Feb. 1989, pp. 10–13.

51. William C. Hsiao, Peter Braun, Douwe Yntema, and Edmund R. Becker, "Estimating physicians' work for a resource-based relative-value scale," *The New England Journal of Medicine*, 1988, vol. 319, pp. 835–841.

52. William L. Roper, "Perspectives on physician-payment reform: the Resource-Based Relative-Value Scale in context," *The New England Journal of Medicine*, 1988, vol. 319, pp. 865–867.

53. The *Journal of the American Medical Association* issue devoted to the Resource-Based Relative-Value Scale was dated Oct. 28, 1988, vol. 260, no. 16.

54. James S. Todd, "At last, a rational way to pay for physicians' services?", *Journal of the American Medical Association*, 1988, vol. 260, p. 2440.

55. Philip R. Lee and Paul B. Ginsburg, "Physician payment reform: an idea whose time has come," *Journal of the American Medical Association*, 1988, vol. 260, p. 2442.

56. In a May 1995 interview, William Hsiao said that the criticism occasionally verged on the vicious. "Once I was addressing the American College of Pathologists," he said, "and I made kind of an off-hand comment like, 'Well, the problem today in American medicine is that physicians don't offer many free services to the patient.' So in this room of about 1,500 people, one pathologist stood up and shouted, 'I would be happy to do a free autopsy on you!' "

57. Ibid.

58. Robert Blendon, personal interview, March 1996.

59. William Hsiao, personal interview, May 1995.

60. Palmer, *Striving for Quality in Health Care,* pp. 5–8.

61. Ibid., p. 7.

62. In 1996 Newhouse was appointed chair, for three years, of the Prospective Payment Assessment Commission, the agency established by Congress in 1984 to monitor and establish hospital rates for Medicare. At that time, diagnosis-related groups (DRGs), a new measure to control hospital costs, were established to set upper limits on what hospitals could expect to be reimbursed from the federal government for caring for Medicare patients.

63. Joseph P. Newhouse, et al., "Some interim results from a controlled trial of cost sharing in health insurance," *The New England Journal of Medicine,* 1981, vol. 305, p. 1501.

64. Joseph P. Newhouse, *Free for All? Lessons from the RAND Health Insurance Experiment.* Cambridge, Mass.: Harvard University Press, 1993, p. 339.

65. See, for example, A. R. Tarlov, et al., "The Medical Outcomes Study: an application of methods for monitoring the results of medical care." *Journal of the American Medical Association,* 1989, vol. 262, pp. 925–930, and Mark Chassin, et al., "Does inappropriate use explain geographic variations in the use of health care services? A study of three procedures." *Journal of the American Medical Association,* 1987, vol. 258, pp. 2533–2537.

66. Robert E. McGarrah, Jr., Patricia Kenney, and Leda R. Judd, "PSRO: doctor accountability or consumer disaster?" In D. Kotelchuck, ed., *Prognosis Negative: Crisis in the Health Care System.* (New York: Vintage Books, 1976, p. 405).

67. Palmer, *Striving for Quality in Health Care,* p. 19.

68. R. Heather Palmer and Mary M. E. Adams, "Quality improvement/quality assurance taxonomy: a framework" in Agency for Health Care Policy and Research, *Putting Research to Work in Quality Improvement and Quality Assurance,* Publication No. 93-0034. Washington, D.C.: U.S. Public Health Service, 1993, p. 13.

69. Palmer, *Striving for Quality in Health Care,* p. 23. According to Palmer, the average American patient is not willing to accept less than the best. "The best," she writes, does not necessarily mean the best in the community or even the state. Increasingly, it has been defined in malpractice suits as nationally established standards of medical practice.

70. Paul C. Weiler, Howard H. Hiatt, Joseph P. Newhouse, William G. Johnson, Troyen A. Brennan, and Lucian L. Leape, *A Measure of Malpractice.* Cambridge, Mass.: Harvard University Press, 1993, p. ix.

71. Ibid., p. xi.

72. Kevin Sack, "More malpractice than lawsuits, New York medical study suggests," *The New York Times,* Jan. 29, 1990, p. 1.

73. Weiler, *A Measure of Malpractice,* p. 55.

74. Ibid., p. 137.

75. Lucian L. Leape, "Error in medicine," *Journal of the American Medical Association,* 1994, vol. 272, p. 1851. Leape cites a study of errors in a medical intensive care unit that showed that medical personnel averaged 1.7 errors per patient per day—out of an average of 178 activities daily. This is only a 1 percent failure rate, but it is higher than what would be tolerated in industry. Even if all systems operated at 99.9 percent success rates, Leape points out, it would be equivalent to two unsafe plane landings per day at Chicago's O'Hare Airport, 16,000 pieces of lost mail every hour, and 32,000 bank checks deducted each hour from the wrong bank account.

76. Ibid., pp. 1852, 1855.

77. Ibid., p. 1857.

78. Howard Raiffa, "A decision analyst's foreword." In *Clinical Decision Analysis,* M. C. Weinstein and H. V. Fineberg, eds. (Philadelphia: W. B. Saunders, 1980, pp. ix–x).

79. Ibid.

80. "Medical care in a national health program. An official statement of the American

Public Health Association adopted October 4, 1944," *American Journal of Public Health,* 1944, vol. 34, pp. 1252–1256.

81. Theodore R. Marmor, "Origins of the government health insurance issue." In D. Kotelchuck, ed., *Prognosis Negative: Crisis in the Health Care System.* New York: Vintage Books, 1976, p. 295.

82. R. Blendon and K. Donelan, "Public opinion and efforts to reform the U.S. health care system: confronting issues of cost-containment and access to care," *Stanford Law and Policy Review,* Fall 1991, p. 147. Cited in Theda Skocpol, "The rise and resounding demise of the Clinton plan," *Health Affairs,* Spring 1995, p. 68.

83. Theda Skocpol, "The rise and resounding demise of the Clinton plan," *Health Affairs,* Spring 1995, p. 67.

84. Robert Blendon, Drew E. Altman, John M. Benson, Humphrey Taylor, Matt Jones, and Mark Smith, "The implications of the 1992 presidential election for health care reform," *Journal of the American Medical Association,* 1992, vol. 268, p. 3371.

85. Marc J. Roberts, with Alexandra T. Clyde. *Your Money or Your Life: The Health Care Crisis Explained.* New York: Doubleday, 1993, pp. 39–40.

86. Ron LaBrecque, "A reader's guide to health care reform," *The Harvard Gazette,* Dec. 3, 1993, p. 14.

87. Robert Blendon, personal interview, March 1994.

88. Robert Blendon, personal interview, March 1996.

89. Roberts, *Your Money or Your Life,* pp. 2–3.

7

The Global Perspective

In reviewing the trials and triumphs of public health over the past 50 years, the inevitable final question is to ponder the future. Although forecasting is a risky proposition, one thing can be predicted with a fair degree of certainty. In the twenty-first century, the overall mission of public health—to protect the public's health by implementing population-wide programs—will remain intact, but the details of carrying out that mission will inevitably change.

The most crucial change will be in the basic orientation toward the social context in which health and disease occur. As Jonathan Mann of the Harvard School of Public Health sees it, the essential difference between "the old public health" of the past 50 years and "the new public health" of today and tomorrow is this:

> The old public health is characterized by a general view that the diseases are dynamic and the society is static, while the new public health considers diseases and society both to be interconnected, so that both are dynamic.[1]

This new view of public health requires that two contemporaneous changes occur: the change to an ever-more interdisciplinary approach and the change to an orientation that is at the same time more global. Once public health experts recognize that society must be as much the focus of study as the diseases themselves, they are forced to turn to colleagues in economics, demography, political science, psychology, sociology, and many other nontraditional disciplines, which all have perspectives to contribute to the study of public health issues. Once they recognize that all of society is global in nature, it is no longer enough to study their own country's social structure; they must see it in the context of every other country's.

205

The most important overarching perspective in contemporary public health is the global perspective. In the twenty-first century, changes or disruptions in one remote corner of the world can reverberate almost at once on a population's health a hemisphere away. With nearly instantaneous communication through satellites and computers, with rapid jet travel bringing everywhere within a day's journey of anywhere, with economic interdependence through multinational monetary policies and expanding international markets, and with health threats such as air pollution and infectious organisms that respect no borders, the world has al-

Jonathan Mann

ready become so interconnected that it no longer makes sense to think of public health in anything but an international way. As Jonathan Mann puts it,

> To talk about community health without reference to national health, to talk about national health without reference to global health, is becoming impossible. To the extent that we understand that economic status is part of the fabric of the social determinants of health status, if the economy is internationally interconnected, so therefore will health be.[2]

This chapter looks at three related trends that can all be taken as signs of a growing globalism in the public health enterprise. The first is the emerging awareness of the link between health and human rights. The promotion and protection of health and the promotion and protection of human rights are inextricably linked. Beyond the complex and dire problems presented by human rights abuses, examination of their relationship to health may well revitalize and change current thinking about public health goals and methods.

The second is population studies, a discipline with a venerable history that has responded to shifts in world population patterns with some shifting orientations of its own. From a focus in the 1950s on birth control dissemination and zero population growth, proponents in the field now focus on the underlying sources of excessive birth rates: poverty, lack of education, and the subjugation of women.

Finally, this chapter looks at international public health and the concept of the "new security" that has grown out of the realignment of power in the wake of the fall of the Soviet Union and its satellites. This global perspective is indeed the future of the field in a nutshell—not only for the sake of America's security but also for the sake of the health of people everywhere.

Health and Human Rights

In 1948 the World Health Organization (WHO) highlighted the intimate connection between the health of a nation and that nation's assurance of basic human rights to all its citizens. In WHO's constitution, health is defined as a state of complete physical, mental, and social well-being. Significantly, WHO goes on to make even more explicit the connection between health and freedom, by stating:

> The enjoyment of the highest attainable standard of health is one of the fundamental rights of every human being without distinction of race, religion, political belief, economic or social condition.[3]

Also in 1948, the United Nations General Assembly adopted the core document of the modern human rights movement, the Universal Declaration of Human Rights. It states, among other things, that all people in all nations have inalienable rights to dignity, security, education, political voice, a decent standard of living, and the chance to raise a family.[4] Significantly, the Universal Declaration is now handed out at commencement ceremonies to all graduates of the Harvard School of Public Health, along with their diplomas. When he presents it to them, the dean says that the document is as central to their futures in public health as the Hippocratic Oath is to new graduates in medicine.[5]

The connection between health and human rights is complex and its connections are only recently being realized. Sometimes, the two forces clash, with the protection of the public's health actually interfering with certain individuals' rights. That is what happened in the 1930s and 1940s when Dr. Fred Lowe Soper tried to eliminate the *Aedes aegypti* mosquito from the Western hemisphere. Soper began in Brazil, where he was invited by President Gutulio Vargas to engage in mosquito abatement programs that would stem the spread of yellow fever. Vargas issued a presidential order allowing Soper and his minions to enter all homes and businesses to inspect for the presence of mosquitoes and to spray insecticides when mosquitoes were found. The aggressive methods worked; *Aedes aegypti* was soon vanquished from Brazilian cities and countryside.

Buoyed by his success, Soper moved north throughout South and Central America, mounting massive eradication campaigns in countries where yellow fever was endemic. But when the mosquito sprayers ran up against the southern border of the United States, where *Aedes aegypti* was also a common pest and yellow fever a frequent visitor, the American tradition of the sanctity of the home put a stop to their efforts. As historian John Duffy put it,

The Global Perspective

207

The result was the gradual reinfiltration of the mosquitoes from the United States into Latin America. In explaining the United States' failure to act, Dr. Soper attributed it to the federal government's lack of jurisdiction and the inability of the American government at any level to enter private premises in search of mosquitoes.[6]

The modern public health community's focus on health and human rights was in large part stimulated by the global confrontation with AIDS and pioneering work by women on women's health issues. One such focus emerged out of AIDS prevention efforts in Uganda. As Jonathan Mann recalls:

> We noted that married, monogamous women in Uganda were increasingly become HIV-infected. At first, we assumed that this was due to ignorance about HIV/AIDS, or to a lack of condom availability. We checked and found that the women knew quite a lot about HIV and its prevention and that condoms were generally available. So, we did what we should have done in the first place—we asked the women![7]

In this way Mann found out that Ugandan women could not refuse the sexual overtures of their husbands, even if the women were uninterested, even if the men were infected with HIV and would not wear condoms. A wife who did not submit risked being beaten (without civil protection or recourse) or being divorced (which meant civil and economic death, since there was no property sharing between spouses). AIDS education was useless in such a setting. The only course of action that had an impact on public health occurred when a group of women lawyers from Uganda proposed reforming the laws governing marriage, inheritance, and property distribution after divorce.

Research interest in health and human rights was significantly enhanced in 1992 by the personal philanthropy of Countess Albina du Boisrouvray. The countess, who was once described as "an unlikely cross between Sonia Braga and Mother Teresa,"[8] was raised in New York, Switzerland, France, and Morocco. The granddaughter of a Bolivian tin tycoon, she worked as a journalist and movie producer after the birth of her only child, François-Xavier Bagnoud, in 1962. The boy grew up to become a rescue pilot, after earning a degree in aerospace engineering from the University of Michigan.

When he was 24 years old, Bagnoud died when his helicopter crashed in Mali. The loss of her son led the countess to establish an association in his memory, built around the theme of rescue in a broad sense. Six years after his death, she endowed an innovative center for health and human rights at the Harvard School of Public Health, and the school named it after her son.

As the center's director and the school's first François-Xavier Bagnoud Professor of Health and Human Rights, Jonathan Mann has sought to define human rights in the broadest possible way. He considers human rights violations to be not only torture or slavery, but far more subtle abuses that would also have significant effects on the public's health.

> What the human rights perspective offers is a coherent framework for societal analysis, a consistent vocabulary for describing situations that look different, and a coherent framework that allows us to identify their commonalities. In addition, the human rights analysis provides clarity about the necessary direction of societal change. Without a coherent framework for analyzing and responding to the societal determinants of health, public health is paralyzed.[9]

"If you think of the underlying cause of a public health problem as poverty," says Mann, "there is a risk that both creative thinking and action stop." The only way to get beyond this paralysis is to take a fresh look at the relationship between poverty and health. Consider famine, for example. While famine is always associated with poverty, in the modern world it never occurs in a country—no matter how poor—that has a free press and some measure of political opposition.[10] "Recognizing the connection between respect for civil and political rights (to information and political participation) and prevention of famine," Mann says, "is part of a growing awareness of the commonalities that link such superficially diverse health issues as breast cancer, child abuse, violence, heart disease, sexually transmitted diseases, drug use, and automobile injuries."[11]

In September 1994 the François-Xavier Bagnoud Center sponsored the First International Conference on Health and Human Rights, to spread information about this new approach. The very occurrence of such a conference, says Mann, indicates a change in the previously stilted dialogue between experts in public health and experts in human rights. "Today, public health workers are beginning to be more aware of the critical relationship between the protection of health and the protection of human rights," he pointed out when the conference began. "Understanding the interaction of these issues will help transform current efforts to ensure a healthy global society."[12]

Tragically, examples of the health impact of human rights violations abound. One of the most dramatic in recent memory is the war in Bosnia, following the 1990 disintegration of the former Yugoslavian republic. After three years of warfare, Bosnia was visited by a fact-finding team from the Harvard/Einstein Study Group, a group of medical and social scientists from the Harvard School of Public Health and the Albert Einstein College of Medicine in New York City. A report of the team's disturbing findings concluded that the war in Bosnia constituted no less than a "war on public health." The statistics from the front were brutal. As of September 1993, the Harvard/Einstein scientists found that, according to some authorities, the conflict had led to the deaths of 200,000 people in the former Yugoslavia, including 30,000 children;

The Global Perspective

209

the rape of 50,000 women; and the dislocation of several million Bosnians. Childhood vaccination rates dropped from 95 percent to 35 percent; rates of diarrhea and dysentery increased substantially between 1992 and 1993, when the water treatment plants in Sarajevo were destroyed by bombing; and malnutrition affected 35 percent of adults in the ravaged city of Srebrenica alone.

"Operations may be performed by candlelight, without anesthesia, by surgeons with freezing hands," the study group wrote. "The hidden dimension of this conflict is an insidious, pernicious assault on the lives and well-being of the people of Bosnia."[13]

The need to look at many public health issues from a human rights perspective has permeated one of the most important issues in global health: population studies. The outcome of the most recent United National World Population Conference on Population and Development, held in Cairo in late 1994, reflects how much the field has changed since it originated in the 1950s. In the early days, population studies meant disseminating birth control and reducing fertility rates in underdeveloped countries. Today, awareness of the connection between health and human rights has changed that orientation. The final document approved in Cairo is, according to Sara Seims of the Rockefeller Foundation,

> among the most forward looking and revolutionary ever to emerge from a meeting of this kind. The conference confirmed the rights of couples and individuals to decide on the number, timing, and spacing of their children; it recognized unsafe abortion as a major public health problem; it confirmed the necessity of integrating contraceptive services as part of a broader program of essential reproductive health interventions; and it linked all of the above with the imperative of improving the status of women and increasing their dignity and empowerment.[14]

The next section charts the evolution of population studies over the past 40 years. When it began, population studies had a rather colonialist orientation: male professionals from the industrialized world brought their contraceptive technologies to female peasants in remote villages and undeveloped cities and tried to limit the number of babies born. In its current form, population studies has become more of a humanitarian discipline. Today it involves male and female health care workers who recognize that it is not just the number of children per woman that determines a society's viability, but the underlying conditions of poverty, ignorance, and inequality that contribute to both overpopulation and the disintegration of public health.

A History of Population Studies

Clarence Gamble, an heir to the Procter & Gamble soap fortune, was a millionaire at the age of 21. He chose to become a scientist and focused his research efforts on contraception. Population control, he wrote, was "the Great Cause."[15] A graduate of Princeton and the Harvard Medical School, Gamble joined the Harvard School of Public Health as a research associate in 1958, by which time he was already 64 years old and had spent more than three decades in contraception research and dissemination. At Harvard, he paid his own salary and he paid for much of his own laboratory space, equipment, and staff as well.

Gamble's first collaboration with the school was his involvement in the India-Harvard-Ludhiana study. Begun in 1953, it was one of the first investigations by an American school of public health in the new field of population science. Gamble sponsored the initial survey in collaboration with epidemiology chairman John E. Gordon, and he provided support for one of his coinvestigators, John Wyon, to attend the Harvard School of Public Health for a year to get a master's degree. Wyon, a British missionary doctor who had been working in rural India, was fluent in Hindi, and after he finished his studies he returned to India to see the study through its first year. Eventually, other groups provided support, including the Rockefeller Foundation, Ford Foundation, and National Science Foundation.

As Gamble's biographers, Doone and Greer Williams, summarize his research, which involved 12,000 Punjabi villagers and a staff of 47 and produced three dozen articles and a book over the course of 17 years:

> The long-term impact of the study from the standpoint of health education as well as contraceptive effectiveness appeared negligible. Its most significant conclusion, and one that provided a guideline for much sociological research to come, was that "Motivation appears to outweigh method in importance." It ultimately proved Gandhi right in one underlying aspect: birth control is a matter of choice—of the will power to do or not to do something.[16]

From a space he rented for himself over Sparr's Drug Store near the Harvard Medical School's Longwood area campus, Gamble conducted research in the late 1950s and early 1960s into the spermicidal properties of salt. While ensconced in these small quarters, he had frequent contact with John C. Snyder, dean of the Harvard School of Public Health since 1954. The two men fed each other's enthusiasm for population control as a central component of the profession.[17]

In 1962, with a generous grant from the Gamble family, Snyder established the Department of Demography and Human Ecology at Harvard. It was the first such department in any school of public health. Within two years, the dean had broadened his vision, creating a university-wide Center for Population Studies. The center encompassed such departments as demography, economics, social science, genetics, education, ethics, and human ecology.[18] By 1967 all candidates for a master's of public health degree at Harvard were required to take a course on population growth and fertility control. Nearly 30 years later,

Snyder considers the population center a prime accomplishment in his deanship and is especially proud of its interdisciplinary approach to a complicated subject. Population studies, he said in 1993, "is not just biochemistry and physiology and pharmacology; it is also the awareness of the fact that people don't change their habits readily, and that introducing social change is an art and a science all of its own."[19]

The irony of population control as a public health goal is that it appears to work at cross purposes with some of the other branches of public health. While most disciplines, such as Snyder's original field of bacteriology, have the effect of *increasing* population by reducing the *death* rate, this one strives to *decrease* population by reducing the *birth* rate. Indeed, as traditional public health programs became more successful—leading to improved living conditions, improved educational attainment, and improved vaccination rates in the developing world—population growth became greater. As Snyder's predecessor as dean, James S. Simmons, noted, "the effective application of public health measures in certain areas will result in increases in the population density of such magnitude that economic and social advances will be vitiated."[20]

Some of Snyder's Egyptian colleagues went so far as to criticize him for his efforts in that country in 1944 to eliminate typhus, telling him: "You're stupid to be doing this. You want the population to be stable, and yet you're trying to enhance the survival of the children."[21]

Snyder's response was that innovations in techniques of education along with changes in the conditions of people in many places in the world would eventually resolve the apparent conflict. In his view the more informed people became about their options, and the more it became apparent to them that the children they bore were likely to make it to adulthood, the more the birth rate would go down. While this has occurred in some populations, progress has not been as rapid as Snyder and his contemporaries had anticipated. "We haven't gotten the dramatic changes in techniques I had hoped would be available by now," he said in 1993. By "techniques" he meant not methods for preventing conception—although those too were in surprisingly short supply—but methods for "motivating people to change their behavior."[22]

Why has putting the brakes on population growth proved so much more difficult than stemming the death rate from infectious disease? In other words, why does the world's population continue to swell despite decades of trying to slow it? One answer was suggested by Roger Revelle, the first director of Harvard's Center for Population Studies. Eliminating disease works according to the laws of nature, he pointed out on the day his appointment was announced in 1964; eliminating births runs directly counter to those natural laws:

> When you try to reduce the death rate, every human instinct is on your side. Everybody wants to live longer. Everybody thinks other people should live longer. When you try to reduce the birth rate, essentially every human instinct works against you. It's a question of

the meaning of life, the question of the Biblical injunction to go forth and multiply. It's the question of the joy of having children and the feeling that you are being a human being if you do have children.[23]

In the years since Revelle's tenure at Harvard, the field of population studies has grown and changed. Although the sheer number of people on the planet is still a major concern—5.7 billion and counting, with an ever-increasing proportion in the poorest nations on earth—population studies today include a greater emphasis on reproductive health. No longer is the goal simply to limit a country's birth rate by setting an absolute number of one or two babies per woman; the goal now is to make it safe for women to have babies when they want, to keep women from having babies when they do not want them, and to help women create a life in which they want fewer children.[24]

The way Revelle conceptualized population studies 30-plus years ago has a distinctly modern ring. Many of his comments are also pertinent to the international perspective of today's public health leaders, especially those who are pushing for a global approach to health for the sake of international harmony and international security. As Revelle said at the time:

> Because of the shrinking size of the world and its growing interdependence, and the fact that all the world's resources are needed to support the world's peoples, an effective way of distributing the world's income more widely among nations must be found if there is to be world prosperity. Without world prosperity there can be no prosperity over any extended period in the U.S.[25]

This is the same orientation that has served public health in the 1990s, as politicians search for a "new world order" in the wake of the fall of the Soviet Union and the realignment of allegiances in the absence of the two Cold War superpowers. From the public health perspective, this change in the global landscape has created a shift in ori-

The Global Perspective

Roger Revelle

213

entation in which the developing nations face an array of economic and social challenges that have led to a new effort in international health reform.

International Public Health and the "New Security"

International security in the wake of the breakup of the Soviet Union is precarious, says Lincoln Chen of the Harvard School of Public Health. Most of the conflicts are intranational rather than international, eruptions of long-standing fissures among different ethnic, religious, or cultural groups within the same country. And the threats to the public's health of these old and bitter conflicts are more pernicious even than the threats of war. As Chen explains:

> Most of the deaths and suffering are inflicted upon civilian populations rather than combatants; indeed, innocent people are often the primary target of conflict rather than mere by-products of war. Threats to human survival and well being are more often silent and invisible, greater than even that reported in the mass media, stemming from the collapse of social and material life support systems. Malnutrition and common infectious diseases, in addition to injury due to violence, are the major causes of death and suffering.[26]

Chen is especially concerned about the health effects of refugee migration, either across national borders or within their country of origin. More than three-quarters of displaced persons are women and children, and death rates can rise up to 30 times normal in the squalid refugee camps into which they are often forced. "The displacement of people powerfully worsens survival risk because hazardous physical circumstances combine with a catastrophic disruption of traditional social support systems—community networks, a family's asset base, and livelihood opportunities," notes Chen. In the Rwandan refugee camps in Zaire, he says, not only do people die at astronomical rates, but they also experience profoundly depressed fertility rates, cessation of marriage, family separation, and social disintegration.[27]

The global response to what he calls "human survival crises" is often unfocused and fails to see beyond the immediate crisis. When ethnic slaughter began in Rwanda in mid-1994, the first response was to provide food supplies to the refugee camps springing up in Zaire. But starvation takes weeks to kill people, Chen points out. In the meantime, clean drinking water and sanitation were much slower to arrive at the camps, even though water-borne infectious diseases "can decimate a population in days."[28]

In June 1994, when the bloodbath in Rwanda was at its peak, the Harvard School of Public Health held its annual commencement. One of the graduating students that day, Vineeta Rastogi, felt compelled to mention in her commencement speech the events taking place half a world away from Boston and to point out the intimate relationship be-

Lincoln Chen

The Global Perspective

tween health and social stability. "Sectarianism is the worst disease we face," Rastogi began. She went on:

> Rwanda saw 200,000 hacked to death in less than a month. No disease is that cruel, that uncaring, that unremitting. Even doctors and nurses who are not killed by the hatred must flee from it. Our work, public health, becomes irrelevant when hate-filled strife becomes the norm.[29]

In the United States the flow of immigrants that has been the backbone of the nation's strength and power has also frequently made the country vulnerable to public health catastrophes of its own. In the 1970s, for instance, as several massive waves of refugees fled their countries because of natural, political, or economic upheavals, they brought with them potentially devastating public health crises. Between 1975 and 1980 the United States saw an influx of 900,000 Southeast Asians (who left their homes in South Vietnam, Cambodia, and Laos after the fall of Saigon in 1975), 125,000 Cubans (most of them prisoners and mental patients fleeing the country in the 1980 Mariel boat lift), and 15,000 Haitians (trying to escape political brutality and economic devastation). Yet because of the efforts of physicians from the U.S. Public Health Service, epidemiologists from the Centers for Disease Control, and state health officers, these tens of thousands of newcomers were all settled in this country without the outbreak of any major disease.[30]

In the 1990s global social changes have occurred even more rapidly, with the result that many nations of the world are undergoing dramatic social and economic upheavals that affect not only the political life of the nation but also the physical health of its people. Socioeconomic changes have led to public health changes in many formerly Communist countries in Central and Eastern Europe; in Communist

215

countries moving toward private markets, such as China and Vietnam; and in the so-called transitional societies—those trying to make the leap from developing to industrialized nations—in Africa and Latin America. In the former Soviet Union, for example, public health officials report the return of cholera, diphtheria, and other once-vanquished infectious diseases. In 1993 one study traced 500,000 excess deaths in Russia alone to changes in the economic marketplace.[31]

The World Bank, which has studied the health care systems of the developing world, reports that in the 1980s and 1990s many developing countries had grossly imbalanced health care budgets, spending $1 a year per capita on such public health measures as immunizations and up to $15 a year per capita on advanced curative medicine. "In some countries," notes Peter Berman of the Harvard School of Public Health, "large teaching hospitals absorb 35 or 40 percent of the entire health budget."[32] During this same period, these countries were undergoing the sort of "health transition" that had taken place in the industrialized world 30 or 40 years before. As infectious diseases yielded to control efforts and child mortality dropped, the population tended to age, so that chronic conditions—such as cancer and heart disease—emerged as the major public health problems. Between 1950 and 1990, child mortality in the developing world dropped from 280 to 106 per 1,000 and life expectancy rose from 40 to 63 years.[33]

For nations in the midst of the health transition, the World Bank laid out a strategy for health reform that focused on three recommendations: (1) improve government health expenditures, with an emphasis on public health programs like AIDS prevention, pollution control, and immunization, and a deemphasis on costly tertiary care facilities and specialist training; (2) promote diversity and competition, so that suppliers of clinical services or drugs submit to market forces to lower prices and improve quality and so that government-regulated private insurance programs are allowed to flourish; and (3) foster an improved health environment, including improved economic policies, improved education, and an improved status for women, an important determinant of a society's overall health status.[34]

The new view of international public health is that it is an important route to a "common security," the name given to an international cooperative research forum organized by the Harvard Center for Population and Development Studies, the Centre for History and Economics at King's College of Cambridge University, and the Norwegian International Center for Social Science Research in Oslo. The main theme of the Common Security Forum is, according to cofounder Lincoln Chen, "that the threats to security in the post–Cold War era have expanded beyond the traditional military protection of national boundaries." The notion of common security is simple, he says: "States can no longer seek security at each other's expense; it can be attained only through cooperative undertakings."[35]

This broader perspective constitutes the future of public health. The new globalism allows for a revitalization of the discipline, for a fresh look at problems that have long stymied efforts to allow the greatest number of people to achieve the highest state of well-being. The mandate of public health, as laid out explicitly by the Institute of Medicine in 1988, is to assure "the conditions in which people can be healthy."[36] To do so, public health scholars in the future will need to see their mission as one that encompasses a wide range of social concerns that go far beyond the traditional boundaries of public health set forth half a century ago. Increasingly, public health must involve not only health ministers and officials but also decision makers responsible for development, economic planning, and welfare programs.

In an era when political and geographic boundaries are permeable, our success in battling threats to human health depends on a clear vision of ourselves as one large, complex community. Many of the threats to public health are international—environmental deterioration, the AIDS epidemic, complex humanitarian emergencies (such as have occurred in Bosnia, Rwanda, and Somalia), and nuclear proliferation—and they demand an international response.[37] Malaria, typhus, and tuberculosis, as well as AIDS; the scourges of poverty, racism, and social isolation; and their particular impact on the elderly, minorities, and children—these are problems that belong to all of us, whether we are directly affected or not. An ongoing responsibility of all public health professionals is to reach those who are the most inaccessible; the well-being of all of us depends on their ability to do so.[38]

Finding shared values, restoring trust and hope—these are the guideposts for public health progress in the United States and around the world. But in a competitive world, how do we find common ground? Sissela Bok, an ethicist and fellow at the Harvard School of Public Health, speaks of the need to develop shared moral values among nations: mutual care and support for family and community members; the censure of violence, lying, and betrayal; and the assertion of certain forms of justice. Such shared values provide basic rules for human conduct and a starting point for deliberation in cases of conflict among or within nations.[39] We can see what happens when such values are shattered by looking at decaying American cities, where young people have lost hope and direct their despair to drugs and violence, or at the shell of the former Yugoslavia, where interethnic rivalries have accounted for a spiraling of atrocities.

Public health as health maker and peace maker—the image is compelling. Organized to attack problems with sound science and analytical methods and renewed by a deeper understanding of the relationship between individuals and society, public health professionals are poised to lead the way to a healthier future. While new threats will inevitably emerge, the essential principles and methods of public health will serve as enduring protection for citizens around the world.

*The Global
Perspective*

Notes

1. Jonathan Mann, personal interview, Oct. 1995.

2. Ibid.

3. "Preamble to the Constitution," *World Health Organization Basic Documents.* Geneva: World Health Organization, 1963, pp. 1–2.

4. The Universal Declaration of Human Rights reads, in part:

"All human beings are born free and equal in dignity and rights. They are endowed with reason and conscience and should act towards one another in a spirit of brotherhood. . . .

"Everyone has a right to life, liberty and the security of person. No one shall be held in slavery or servitude; slavery and the slave trade shall be prohibited in all their forms. No one shall be subjected to torture or to cruel, inhuman or degrading treatment or punishment. Everyone has the right to recognition everywhere as a person before the law. . . .

"Men and women of full age, without any limitation due to race, nationality or religion, have the right to marry and to found a family. . . . The family is the natural and fundamental group unit of society and is entitled to protection by society and the State. . . .

"Everyone has the right to freedom of thought, conscience and religion; . . . to freedom of opinion and expression; . . . to freedom of peaceful assembly and association. . . . Everyone, as a member of society, has the right to social security and is entitled to realization . . . of the economic, social and cultural rights indispensable for his dignity and the free development of his personality.

"Everyone has the right to work, to free choice of employment, to just and favourable conditions of work and to protection against unemployment. Everyone, without any discrimination, has the right to equal pay for equal work. . . Everyone has the right to rest and leisure, including reasonable limitation of working hours and periodic holidays with pay.

"Everyone has the right to a standard of living adequate for the health and well-being of himself and of his family, including food, clothing, housing and medical care and necessary social services, and the right to security in the event of unemployment, sickness, disability, widowhood, old age or other lack of livelihood in circumstances beyond his control. Motherhood and childhood are entitled to special care and assistance. All children, whether born in or out of wedlock, shall enjoy the same social protection.

"Everyone has the right to education. . . . Education shall be directed to the full development of the human personality and to the strengthening of respect for human rights and fundamental freedoms. It shall promote understanding, tolerance and friendship among all nations, racial or religious groups . . . Everyone has the right [to freely] participate in the cultural life of the community, to enjoy the arts and to share in scientific advancement and its benefits."

5. According to Mann (personal interview, Oct. 1995), this has been the practice at the Harvard School of Public Health commencement ceremonies since 1990.

6. John Duffy, *The Sanitarians: A History of American Public Health.* Urbana: University of Illinois Press, 1990, p. 4.

7. Jonathan Mann, personal interview, Feb. 1996.

8. Michael Z. Wise, "The countess's golden touch," *The Washington Post,* Dec. 8, 1992, p. D1.

9. Jonathan Mann, personal interview, Oct. 1995.

10. The relationship between famine and political oppression has been observed by Amartya Sen in many of his writings. Sen is Harvard Lamont University professor in the economics department. See Amartya Sen, "The economics of life and death," *Scientific American,* May 1993, p. 43, and "Individual freedom as a social commitment," Amartya Kumar Sen's address at the award ceremony of the Senator Giovanni Agnelli International Prize, Turin, March 5, 1990.

11. Jonathan Mann, "Public health and human rights," *Current Issues in Public Health,* 1995, vol. 1, p. 100.

12. "Conference on health and human rights," *Around the School: News & Notices of the Harvard School of Public Health,* Sept. 16, 1994, p. 1.

13. "Bosnia + Haiti: crisis in humanitarian action." In *Ideas in Action: 1994 Annual Report.* Boston: Harvard School of Public Health, 1994, p. 13.

14. Sara Seims, "Population and development: an overview and perspective of the Cairo conference," *Current Issues in Public Health,* 1995, vol. 1, pp. 42–48.

15. Doone and Greer Williams, with Emily P. Flint, *Every Child a Wanted Child: Clarence James Gamble, M.D., and His Work in the Birth Control Movement.* Boston: The Francis A. Countway Library of Medicine, 1978.

16. Ibid., p. 188.

17. Early influences on Snyder concerning the problem of population growth occurred during his tenure in the early 1940s with the Rockefeller Foundation in Mexico and Spain and with the USA Typhus Commission in the Middle East and Europe. When he joined Harvard in 1946, subsequent contacts with students from developing countries, collaborations with Harvard faculty members John Gordon, Carl Taylor, and John Wyon, and consultations with contraception pioneer John Rock gave him added incentive to pursue an understanding of the complex forces affecting world population growth.

18. One of the reasons for the success of the interdisciplinary Harvard Center for Population Studies, one of the first of its kind, was the leadership of its director, Roger Revelle. Revelle had no background in public health or in population studies—when he was recruited to Harvard, he had spent 33 years at the University of California, the last 13 as director of the Scripps Institution of Oceanography—but he took a broad approach to the subject. (And, perhaps not coincidentally, he was married to a classmate of John Snyder's from Pasadena High School—an added personal inducement to accept the appointment.) "Although Oceanography would not seem to fit categorically into human population studies," the school's public relations officer wrote in the *Harvard Public Health Alumni Bulletin* (July 1964, vol. 21, p. 12), "Dr. Revelle's interests and competencies are clearly not bound by arbitrary disciplinary boundaries."

19. John Snyder, personal interview, Nov. 1993.

20. "New department will deal with population problems," *Harvard Public Health Alumni Bulletin,* Dec. 1962, vol. 19, p. 26.

21. Snyder, personal interview.

22. Ibid.

23. "The University Population Center: Excerpts from press conference with Dr. Roger Revelle, director of the Harvard Center for Population Studies, October 5, 1964," *Harvard Public Health Alumni Bulletin,* Jan. 1965, vol. 22, p. 14.

24. "Women's roles emphasized in program on world population," *Around the School: News & Notices of the Harvard School of Public Health,* Oct. 27, 1995, p. 3.

25. Excerpts from press conference with Dr. Roger Revelle," *Harvard Public Health Alumni Bulletin.*

26. Lincoln C. Chen and Aafje Rietveld, "Human security during complex humanitarian emergencies: rapid assessment and institutional capabilities," *Medicine & Global Survival,* 1994, vol. 1, p. 156. The paper was originally delivered at a workshop on "New Challenges for Humanitarian Intervention" sponsored by the Program on Human Survival and Security of the Harvard Center for Population and Development Studies, which was held May 14–15, 1994, at the Harvard Divinity School.

27. Ibid., pp. 156–157.

28. Ibid., p. 157.

29. Vineeta Rastogi, commencement speech, cited in *Around the School: News & Notices of the Harvard School of Public Health,* June 17, 1994, p. 2. Eighteen months after giving her speech, Rastogi died of cancer at the age of 26. A graduate of the University of Maryland who spent time in Zaire with the Peace Corps, she was memorialized by her

former professor Colman McCarthy (in *The Washington Post*, Dec. 12, 1995) as "everyone's friend, a giver, a listener and someone whose subterranean thirst for social justice prompted those around her to push themselves to become better people, as she was doing." During her two years at Harvard, she bicycled through Vietnam with her fiancé, Brian Hennessey, visiting clinics and hospitals en route, and also worked with health care workers in El Salvador, Cuba, and India. She was, wrote McCarthy, "one of the rare ones who aimed higher than high . . . Vineeta knew what she owed. And public health, one of the highest callings of all, was her plan to pay in full."

30. Fitzhugh Mullan, *Plagues and Politics: The Story of the United States Public Health Service*. New York, Basic Books, 1989, pp. 191–192. Mullan points out that international public health officials also were dispatched routinely throughout the 1970s to hotspots around the globe that could breed epidemics: to Nigeria during its 1967–1970 civil war; to Guatemala after the 1976 earthquake; to sub-Saharan Africa during the 1973–1975 famine years. These efforts included the investigation of several bizarre outbreaks of deadly hemorrhagic fevers in Africa and Europe as well: Marburg fever in Germany in 1976; Lassa fever in Sierra Leone, Sudan, and Zaire in 1976; and Ebola virus in Zaire and Sudan in 1976.

31. UNICEF, *Public Policy and Social Conditions: Central and Eastern Europe in Transition*. Florence: UNICEF International Child Development Centre, 1993.

32. Richard Anthony, "Blueprint for change: developing countries are placing a new emphasis on health care reform," *Harvard Public Health Review*, Spring 1994, p. 24.

33. Ibid., p. 25.

34. Ibid.

35. Chen and Rietveld, "Human security during complex humanitarian emergencies," p. 162.

36. Institute of Medicine, Committee for the Study of the Future of Public Health, *The Future of Public Health*. Washington, D.C.: National Academy Press, 1988, p. 7.

37. Sissela Bok, "Cultural diversity and shared moral values." In T. Matsumae and L. C. Chen, eds., *Common Security in Asia: New Concepts of Human Security*. Tokyo: Tokai University Press, 1995, p. 27.

38. Harvey V. Fineberg and Mary E. Wilson, "Social vulnerability and death by infection," *The New England Journal of Medicine*, 1996, vol. 334, p. 860.

39. Bok, "Cultural diversity and shared moral values," p. 22.

Acknowledgments

This "institutional memoir" has been enlivened by the insights of dozens of faculty members from the Harvard School of Public Health, as well as a few outside observers from elsewhere in the Harvard or public health communities. For the sake of brevity, their full titles (which can be enormously cumbersome) were eliminated from the body of the text. But I want to give them their full titles here—and to thank them for the time they spent with me, both sitting for interviews and helping me get the details of the mansucript precisely right. My thanks, too, to Bernita Anderson, Elizabeth Jones, and Peter Wehrwein for their careful proofreading and editing; to Beverly Freeman, director of public affairs at the Harvard School of Public Health, for her consistent and uncomplaining help throughout; and especially to Dean Harvey V. Fineberg, for conceiving of this project in the first place and for his gracious guidance as together we carried it to completion. As always, I bear full responsibility for any errors that have slipped through our painstaking fact-checking process, but I recognize that I have been saved from embarrassment several times because of the cooperation of the good people listed below.

Lisa F. Berkman, MS, PhD, Florence Sprague Norman and Laura Smart Norman Professor of Health and Social Behavior and Epidemiology, and Chair of the Department of Health and Social Behavior, Harvard School of Public Health

Peter A. Berman, MSc, PhD, Associate Professor of International Health Economics, Harvard School of Public Health

Donald M. Berwick, MD, MPP, Adjunct Associate Professor of Health Policy and Management, Harvard School of Public Health, and President and CEO, Institute for Healthcare Improvement, Boston

Robert J. Blendon, MBA, MPH, DSc, Professor of Health Policy and Management, Harvard School of Public Health

Joseph D. Brain, SM, SM in Hyg, SD in Hyg, Cecil K. and Philip Drinker Professor of Environmental Physiology, and Chair of the Department of Environmental Health, Harvard School of Public Health

Lincoln C. Chen, MD, MPH, Taro Takemi Professor of International Health, Chair of the Department of Population and International Health, Harvard School of Public Health, and Director of the Harvard Center for Population and Development Studies

Graham A. Colditz, MBBS, MPH, DPH, Associate Professor in the Department of Epidemiology, Harvard School of Public Health, and Associate Professor of Medicine, Harvard Medical School

John R. David, MD, Richard Pearson Strong Professor of Tropical Public Health, and Chair of the Department of Tropical Public Health, Harvard School of Public Health, and Professor of Medicine, Harvard Medical School

Douglas W. Dockery, SM, SM, SD, Associate Professor of Environmental Epidemiology, Harvard School of Public Health, and Associate Professor of Medicine, Harvard Medical School

Felton J. (Tony) Earls, MD, Professor of Human Behavior and Development, Harvard School of Public Health, and Professor of Child Psychiatry, Harvard Medical School

Myron E. (Max) Essex, DVM, PhD, Mary Woodard Lasker Professor of Health Sciences, Chair of the Department of Cancer Biology, Harvard School of Public Health, and Chairman of the Harvard AIDS Institute

Harvey V. Fineberg, MD, MPP, PhD, Dean of the Faculty of Public Health, and Professor of Health Policy and Management, Harvard School of Public Health

William Foege, MD, MPH, Executive Director of the Task Force for Child Survival and Development, Emory University (also, former director of the Centers for Disease Control, and a 1965 alumnus of the Harvard School of Public Health)

Howard Frazier, MD, Professor of Health Policy and Management, *Emeritus*, Harvard School of Public Health, and Professor of Medicine, *Emeritus*, Harvard Medical School

John D. Graham, PhD, Professor of Policy and Decision Sciences, Harvard School of Public Health, Director of the Center for Risk Analysis, and Director of the Harvard Injury Control Center

Gareth M. Green, MD, Professor of Environmental Health and Associate Dean for Professional Education, Harvard School of Public Health

Joseph J. Harrington, AM, PhD, Professor of Environmental Health Engineering, Harvard School of Public Health, and Gordon McKay Professor of Environmental Engineering, Harvard University

Charles H. Hennekens, MD, MPH, MS, DrPH, Professor in the Department of Epidemiology, Harvard School of Public Health, and John Snow Professor of Medicine, Harvard Medical School

Howard H. Hiatt, MD, Professor of Medicine, Harvard Medical School

William C. Hsiao, MPA, PhD, K.T. Li Professor of Economics, Harvard School of Public Health

Howard Hu, MD, MPH, SM, ScD, Associate Professor of Occupational Medicine, Harvard School of Public Health, and Associate Professor of Medicine, Harvard Medical School

Camara P. Jones, MD, PhD, Assistant Professor of Health and Social Behavior, and Assistant Professor of Epidemiology, Harvard School of Public Health

Phyllis Kanki, DVM, SD, Associate Professor of Pathobiology, Harvard School of Public Health

Stephen W. Lagakos, MPhil, PhD, Professor of Biostatistics, Harvard School of Public Health

Nan M. Laird, PhD, Henry Pickering Walcott Professor of Biostatistics, and Chair of the Department of Biostatistics, Harvard School of Public Health

Lucian L. Leape, MD, Adjunct Professor of Health Policy, Harvard School of Public Health

Sol Levine, MA, PhD, Professor of Health and Social Behavior, Harvard School of Public Health, and Senior Scientist, New England Medical Center

John B. Little, MD, James Stevens Simmons Professor of Radiobiology, Harvard School of Public Health

Bernard Lown, MD, DPH, PhD, SM, DSc (Hon.), Professor of Cardiology in Nutrition, *Emeritus*, Harvard School of Public Health

Adetokunbo Lucas, MD, SM, Professor of International Health, *Emeritus*, Harvard School of Public Health

Brian MacMahon, SM, MD, DPH, PhD, Henry Pickering Walcott Professor of Epidemiology, *Emeritus*, Harvard School of Public Health

Jonathan Mann, MPH, MD, François-Xavier Bagnoud Professor of Health and Human Rights, Professor of Epidemiology and International Health, and Director of the François-Xavier Bagnoud Center for Health and Human Rights, Harvard School of Public Health

Marie C. McCormick, MD, ScD, Professor of Maternal and Child Health, and Chair of the Department of Maternal and Child Health, Harvard School of Public Health, and Professor of Pediatrics, Harvard Medical School

Dade W. Moeller, PhD, AM (Hon.), Professor of Engineering and Environmental Health, *Emeritus*, Harvard School of Public Health

Richard Monson, MD, SM, SD, Professor of Epidemiology, Harvard School of Public Health

Frederick Mosteller, PhD, SD (Hon.), SSD (Hon.) LLD (Hon.), Roger Irving Lee Professor of Mathematical Statistics, *Emeritus*, Harvard School of Public Health

Nancy E. Mueller, SM, SD, Professor of Epidemiology, Harvard School of Public Health

Joseph P. Newhouse, PhD, PhD, John D. MacArthur Professor of Health Policy and Management in the Faculties of Medicine, Government, Public Health, and Arts and Sciences, and Director of the Division of Health Policy Research and Education, Harvard University

Ann R. Oliver, Associate Dean for Academic Affairs, Harvard School of Public Health

R. Heather Palmer, MB, BCh, SM in Hyg, Lecturer on Health Services, and Director of the Harvard Center for Quality of Care Research and Education, Harvard School of Public Health

Deborah B. Prothrow-Stith, MD, Assistant Dean for Government and Community Programs, and Professor of Public Health Practice, Harvard School of Public Health

Marc J. Roberts, PhD, Professor of Political Economy, Kennedy School of Government and Harvard School of Public Health

Barbara G. Rosenkrantz, PhD, Professor of the History of Science, *Emerita*, Harvard University

John C. Snyder, MD, LLD, Professor of Population and Public Health, *Emeritus*, and former Dean, Harvard School of Public Health

Andrew Spielman, ScD, Professor of Tropical Public Health, Harvard School of Public Health

Meir Stampfer, MD, MPH, DPH, Professor of Epidemiology and Nutrition, Harvard School of Public Health

Fredrick J. Stare, PhD, MD, SD (Hon.), DSc (Hon.), Professor of Nutrition, *Emeritus*, Harvard School of Public Health

Armen H. Tashjian, Jr., MD, Professor of Toxicology and Chair of the Department of Molecular and Cellular Toxicology, Harvard School of Public Health, and Professor of Biological Chemistry and Molecular Pharmacology, Harvard Medical School

Dimitrios Trichopoulos, MD, Vincent L. Gregory Professor of Cancer Prevention, Harvard School of Public Health

Henry Wechsler, AM, PhD, Lecturer on Social Psychology, and Director, College Alcohol Studies Project, Harvard School of Public Health

Milton C. Weinstein, AM, MPP, PhD, Henry J. Kaiser Professor of Health Policy and Management, Harvard School of Public Health

James L. Whittenberger, MD, AM (Hon.), James Stevens Simmons Professor of Public Health and Physiology, *Emeritus*, Harvard School of Public Health

Walter C. Willett, MD, MPH, DrPH, Fredrick John Stare Professor of Epidemiology and Nutrition, Chair of the Department of Nutrition, Harvard School of Public Health, and Professor of Medicine, Harvard Medical School

Mary E. Wilson, MD, Assistant Professor of Population and International Health, and Assistant Professor of Epidemiology, Harvard School of

Public Health; Assistant Clinical Professor of Medicine, Harvard Medical School; and Chief of Infectious Diseases, Mount Auburn Hospital

Jay A. Winsten, PhD, Associate Dean for Public and Community Affairs, and Director of the Center for Health Communication, Harvard School of Public Health

Dyann F. Wirth, PhD, Professor of Tropical Public Health, Harvard School of Public Health

Marvin Zelen, AM, PhD, Professor of Statistical Science, Harvard School of Public Health, and Director of Statistics Unit, Dana-Farber Cancer Institute

Acknowledgments

225

Bibliography

Ackerknecht, Erwin H. *A Short History of Medicine,* revised edition. Baltimore and London: The Johns Hopkins University Press, 1982.

Bayer, Ronald. *Private Acts, Social Consequences: AIDS and the Politics of Public Health.* New York: The Free Press, 1989.

Berkman, Lisa F. and Lester Breslow. *Health and Ways of Living: The Alameda County Study.* New York and Oxford: Oxford University Press, 1983.

Bowen, Catherine Drinker. *Family Portrait.* Boston: Little, Brown, 1970.

Bunker, John R., Benjamin A. Barnes, and Frederick Mosteller. *Costs, Risks, and Benefits of Surgery.* New York: Oxford University Press, 1977.

California Medical Association and California Hospital Association. *Report on the Medical Insurance Feasibility Study.* San Francisco: Sutter Publications, 1977.

Chernin, Eli. *Tropical Medicine at Harvard: The Weller Years, 1954-1981. A Personal Memoir.* Boston: Harvard School of Public Health, 1985.

Cohn, Victor. *News & Numbers: A Guide to Reporting Statistical Claims and Controversies in Health and Related Fields.* Ames, Iowa: Iowa State University Press, 1989.

Crain, Robert L., Elihu Katz, and Donald B. Rosenthal. *The Politics of Community Conflict: The Fluoridation Decision.* Indianapolis and New York: The Bobbs-Merrill Company, 1969.

Curran, Jean Alonzo. *Founders of the Harvard School of Public Health: With Biographical Notes, 1909-1946.* New York: Josiah Macy, Jr., Foundation, 1970.

Danzon, Patricia. *Medical Malpractice: Theory, Evidence and Public Policy.* Cambridge, MA: Harvard University Press, 1985.

227

Desowitz, Robert S. *The Malaria Capers: More Tales of Parasites and People, Research and Reality.* New York and London: W. W. Norton, 1991.

Diethrich, Edward B. and Carol Cohan, *Women and Heart Disease.* New York: Times Books, 1992.

Duffy, John. *The Sanitarians: A History of American Public Health.* Urbana and Chicago: University of Illinois Press, 1990.

Dutton, Diana B. *Worse Than the Disease: Pitfalls of Medical Progress.* Cambridge and New York: Cambridge University Press, 1988.

Ehrenreich, Barbara and John Ehrenreich. *The American Health Empire: Power, Profits and Politics.* New York: Vintage Books, 1970.

Fee, Elizabeth and Roy M. Acheson (eds). *A History of Education in Public Health: Health That Mocks the Doctors' Rules.* Oxford and New York: Oxford University Press, 1981.

Garrett, Laurie. *The Coming Plague: Newly Emerging Diseases in a World out of Balance.* New York: Farrar, Strauss and Giroux, 1994.

Health and Human Services, Department of. *Healthy People 2000: National Health Promotion and Disease Prevention Objectives.* Washington, D.C.: Public Health Service, 1991.

Henig, Robin Marantz. *A Dancing Matrix: Voyages Along the Viral Frontier.* New York: Alfred A. Knopf, 1993.

Hunt, Vilma R. *Work and the Health of Women.* Boca Raton, Fla.: CRC Press, 1979.

Institute of Medicine. *The Future of Public Health.* Washington, D.C.: National Academy Press, 1988.

Kotelchuck, David (ed). *Prognosis Negative: Crisis in the Health Care System.* New York: Vintage Books, 1976.

Leavitt, Judith Walzer and Ronald L. Numbers (eds). *Sickness & Health in America: Readings in the History of Medicine and Public Health (2nd edition).* Madison and London: The University of Wisconsin Press, 1985.

Leighton, Alexander H. *My Name is Legion: Foundations for a Theory of Man in Relation to Culture.* New York: Basic Books, 1959.

Lewis, Sinclair. *Arrowsmith.* New York: Harcourt, Brace, 1925.

Mack, Arien (ed), *In Time of Plague: The History and Social Consequences of Lethal Epidemic Disease.* New York and London: New York University Press, 1991.

Meyers, Robert. *DES: The Bitter Pill.* New York: Seaview/Putnam, 1983.

Mullan, Fitzhugh. *Plagues and Politics: The Story of the United States Public Health Service.* New York: Basic Books, 1989.

Neustadt, Richard E. and Harvey V. Fineberg. *The Swine Flu Affair: Decision-Making on a Slippery Slope.* Washington, D.C.: Department of Health, Education, and Welfare, 1978.

Newhouse, Joseph P. *Free for All? Lessons from the RAND Health Insurance Experiment.* Cambridge, Mass.: Harvard University Press, 1993.

Nikiforuk, Andrew. *The Fourth Horseman: A Short History of Epidemics,*

Plagues, Famines, & Other Scourges. New York: M. Evand & Company, 1991.

Norwood, Christopher. *At Highest Risk: Protecting Children from Environmental Injury.* New York: Penguin Books, 1981.

Paul, Benjamin D., ed. *Health, Culture and Community: Case Studies of Public Reactions to Health Programs.* New York: Russell Sage Foundation, 1955.

Prothrow-Stith, Deborah and Michaele Weissman. *Deadly Consequences.* New York: HarperCollins, 1991.

Roberts, Marc J. and Alexandra T. Clyde. *Your Money or Your Life: The Health Care Crisis Explained.* New York: Doubleday, 1993.

Rosen, George. *A History of Public Health (expanded edition).* Baltimore and London: The Johns Hopkins University Press, 1991.

Russell, Louise B. *Is Prevention Better Than Cure?* Washington, D.C.: Brookings Institution, 1986.

Shorter, Edward. *The Health Century.* New York: Doubleday, 1987.

Sicherman, Barbara. *Alice Hamilton: A Life in Letters.* Cambridge, Mass.: Harvard University Press, 1984.

Sidel, Victor W. and Ruth Sidel (eds.). *Reforming Medicine: Lessons of the Last Quarter Century.* New York: Pantheon Books, 1984.

Stare, Fredrick J. *Adventures in Nutrition.* Hanover, Mass.: The Christopher Publishing House, 1991.

Starr, Paul. *The Social Transformation of American Medicine.* New York: Basic Books, 1982.

Thomas, Lewis. *The Youngest Science: Notes of a Medicine-Watcher.* New York: The Viking Press, 1983.

Weiler, Paul C., Howard H. Hiatt, Joseph P. Newhouse, William G. Johnson, Troyen A. Brennan, and Lucian L. Leape. *A Measure of Malpractice: Medical Injury, Malpractice Litigation, and Patient Compensation.* Cambridge, Mass.: Harvard University Press, 1993.

Weinstein, Milton C. and Harvey V. Fineberg, with Arthur S. Elstein, Howard S. Frazier, Duncan Neuhauser, Raymond R. Neutra, and Barbara J. McNeil. *Clinical Decision Analysis.* Philadelphia, London, and Toronto: W.B. Saunders Company, 1980.

Weinstein, Milton C. and William B. Stason. *Hypertension: A Policy Perspective.* Cambridge, Mass. and London: Harvard University Press, 1976.

Weisse, Allen B. *Medical Odysseys: The Different and Sometimes Unexpected Pathways to Twentieth-Century Medical Discoveries.* New Brunswick, N.J.: Rutgers University Press, 1991.

Williams, Doone and Greer Williams, with Emily P. Flint. *Every Child a Wanted Child: Clarence James Gamble, M.D., and His Work in the Birth Control Movement.* Boston: The Francis A. Countway Library of Medicine, 1978.

Williams, Greer. *Virus Hunters.* New York: Alfred A. Knopf, 1959.

Photo Credits

With the exception of the items listed below, all the photographs and illustrations in this volume are courtesy of the Harvard School of Public Health, which has granted permission for their use.

Page	Credit
4	Richard A. Chase
10 (bottom left)	Christopher Morrow
10 (bottom right)	Richard A. Chase
35	Jeff Bach
38	Centre National de la Recherche Scientifique
39	Larry Maglott
48	Kent Dayton
69	Kent Dayton
71	Richard A. Chase
94	Betsy Cullen
100	Kent Dayton
107	Paula Lerner
110	Kent Dayton
135	Kent Dayton
146	Richard A. Chase
149	Richard A. Chase
156	Lisa Berg
165	Sue Owrutsky
174	Richard A. Chase
178	Kent Dayton
180	Sue Owrutsky
183	Kent Dayton
189	Richard A. Chase
198	Richard A. Chase
215	Betsy Cullen

About the Author

Robin Marantz Henig is a freelance writer who interprets complex medical, psychological, and health policy issues for an informed lay readership. She has written five books, most recently *A Dancing Matrix: How Science Confronts Emerging Viruses*, for which she was named the 1994 Author of the Year of the American Society of Journalists and Authors. Her articles about health and medicine have appeared in numerous national publications, including *The New York Times Magazine, American Health, Civilization, Woman's Day, Ms., Self, Vogue, Mirabella*, and *The Washingtonian*. Her news and op-ed articles have appeared in *The New York Times* and *The Washington Post*, and she has written dozens of book reviews for both newspapers.

A native of New York City, Henig now lives in Takoma Park, Maryland, with her husband, Jeff, a professor of political science at George Washington University, and their teenaged daughters, Jessica and Samantha. She majored in English at Cornell University, and earned a master's degree in journalism at Northwestern. She is past president of the D.C. chapter of the American Society of Journalists and Authors, and is also a member of the National Association of Science Writers and The Authors Guild.

Index

B

C

Index

Index

Index